AF352977

New Indicators for the Assessment and Prevention of Noise Nuisance

New Indicators for the Assessment and Prevention of Noise Nuisance

Editors

Luca Fredianelli
Peter Lercher
Gaetano Licitra

MDPI • Basel • Beijing • Wuhan • Barcelona • Belgrade • Manchester • Tokyo • Cluj • Tianjin

Editors
Luca Fredianelli
National Research Council,
Institute of Chemical and
Physical Processes
Italy

Peter Lercher
Graz University of
Technology
Austria

Gaetano Licitra
Environmental Protection
Agency of Tuscany Region
Italy

Editorial Office
MDPI
St. Alban-Anlage 66
4052 Basel, Switzerland

This is a reprint of articles from the Special Issue published online in the open access journal *International Journal of Environmental Research and Public Health* (ISSN 1660-4601) (available at: https://www.mdpi.com/journal/ijerph/special_issues/noise_nuisance).

For citation purposes, cite each article independently as indicated on the article page online and as indicated below:

LastName, A.A.; LastName, B.B.; LastName, C.C. Article Title. *Journal Name* **Year**, *Volume Number*, Page Range.

ISBN 978-3-0365-5739-7 (Hbk)
ISBN 978-3-0365-5740-3 (PDF)

Contents

About the Editors

Luca Fredianelli

Luca Fredianelli (Ph.D., Adjunct Professor) Full-time researcher at the Italian National Council of Research (CNR) and qualified associate professor in Applied Physics, Physics Teaching and History Of Physics—FIS/07 and adjunct professor for the University of Pisa. He received his Ph.D in Environmental Physics, and has an active interest in the field of acoustics in all its aspects and in particular in Environmental Acoustics. He has carried out research through collaboration on numerous national and international projects. His research has appeared in more than 20 publications in indexed international scientific journals and about 30 conference proceedings, as well as chapters in books. He has presented his work at numerous national and international conferences. He leads acoustics courses for professionals required by law for Acoustical Technicians, and he has worked in teaching support at the University. Prof. Fradianelli has acted as a reviewer and editor for multiple journals. With more than 10 years of professional experience, he works within the acoustics sector to carry out action plans, acoustic zoning, noise maps and various acoustic impact assessments.

Peter Lercher

Peter Lercher (Professor, Retired) has a background in general medicine, hygiene, social medicine and environmental health. He worked as Associate Professor for Social Medicine at the Medical University of Innsbruck and since 2010 as Director of the Division of Social Medicine.

Dr. Lercher served as chair/co-chair (1993–1998; 1998–2003) of Team 3 in the International Commission on the Biological Effects of Noise (ICBEN) and was Co-Chair of ICBEN between 2003 and 2008.

He was Primary investigator in several multi-method studies related to integrated assessment in EHIA ("Lower Inn valley-rail" and "BBT" (Brenner Base Tunnel), the ALPNAP-project, which aimed at better predictions of noise and air pollution exposure and health effects due to transportation in alpine areas.

He was special adviser in the EU-project ENNAH (European Network on Noise and Health) and Working group chair in the COST action TD 0804 "Soundscape of European Cities and Landscapes" (2009–2013). Afterwards, he worked in the WHO Noise guideline development group and was editor of the special issue "WHO Noise and Health Evidence Reviews" in the IJERPH in 2018 and PI "Alternative noise indicators for health studies" (2016–2018). After his retirement—since 2018—associated with TU Graz and currently PI in the Horizon 2020 funded project "Equal-Life" (2020–2024).

Gaetano Licitra

Gaetano Licitra (Professor) BSc and MSc Physics Dipl. Sp. Health Physics—Head of the Provincial Department of Pisa of the Environmental Protection Agency of the Tuscany Region (ARPAT). Associated researcher at IPCF-CNR and adjunct Professor at the University of Pisa. Author of more than 300 publications https://orcid.org/0000-0003-4867-0954. He has more than 25 years' experience in noise management. He was the scientific reference for ARPAT of about 10 EU funded projects on noise and was appointed by ARPAT as LEAR. He was entitled temporary TAIEX expert of the EC on the actions for the implementation of the Directive 49/2002/EC in Lithuania, Cyprus, Serbia, Macedonia, Kosovo. In June 2022, he hosted TAIEX-EIR event 80087 on Noise protection with Poland. He was "Not key expert" in EuropeAid/131352/D/SER/TR. Editor and reviewer of many journals.

International Journal of
Environmental Research and Public Health

Editorial

New Indicators for the Assessment and Prevention of Noise Nuisance

Luca Fredianelli [1,*]**, Peter Lercher** [2] **and Gaetano Licitra** [1,3,*]

[1] Institute of Chemical and Physical Processes of National Research Council, Via G. Moruzzi 1, 56124 Pisa, Italy
[2] Institute for Highway Engineering and Transport Planning of Graz University of Technology, Rechbauerstraße 12/II, 8010 Graz, Austria
[3] Environmental Protection Agency of Tuscany Region, Via Vittorio Veneto 27, 56127 Pisa, Italy
* Correspondence: luca.fredianelli@ipcf.cnr.it (L.F.); g.licitra@arpat.toscana.it (G.L.)

Keywords: noise indicators; noise metrics; psychoacoustic; nuisance; annoyance; sleep disturbance; peak noise; impulsive events; health related quality of life; health effects

Citation: Fredianelli, L.; Lercher, P.; Licitra, G. New Indicators for the Assessment and Prevention of Noise Nuisance. *Int. J. Environ. Res. Public Health* **2022**, *19*, 12724. https://doi.org/10.3390/ijerph191912724

Received: 29 September 2022
Accepted: 30 September 2022
Published: 5 October 2022

Publisher's Note: MDPI stays neutral with regard to jurisdictional claims in published maps and institutional affiliations.

1. Introduction

At present, health effects induced by prolonged noise exposure are widely studied to determine the most spread noise sources and their effects. Environmental epidemiology studies have mostly associated long-term exposure to high noise levels (>85 dB) in occupational (military, construction workers, and agriculture) and leisure settings with direct auditory effects, including hearing loss [1], tinnitus, and hyperacusis [2], whereas non-auditory effects are limited [3] due to occupational selection effects. On the other hand, long-term exposure to low–medium levels (45–65 dB) in general population settings is followed by a broad spectrum of non-auditory health effects [4,5]. Annoyance [6], sleep disturbance [7], cognitive impairment [8], behavioral and emotional disorders in children and adolescents [9], depression and anxiety in adults [10], stronger physiological stress reactions [11], and endocrine imbalance and cardiovascular disorders [12] are evidence-based effects. Overall, noise is among the leading risk factors for cardiovascular morbidity and mortality worldwide [13]. Usually, these studies regress the health effects of noise on acoustic exposure metrics, which is the average energetic dose over a long time period, such as equivalent continuous sound pressure level (L_{eq}), with its A-weighted version L_{Aeq}, or the day–evening–night level (L_{den}).

Only recently, the scientific community has realized that health effect studies of long-term noise exposure should consider a broader spectrum of sound exposure features as well. Among those features are its intensity variation over time, the impulsivity of events, the frequency distribution, and psychoacoustics parameters. Peak levels, maximum levels, and variability can have a significative influence on nuisance perception, and citizens are known to complain more about single high levels rather than average exposure. This can be even more important for non-traffic-related sources such as leisure noise [2], or even less investigated sound sources such as ships [14] or wind turbines [15]. Generally, an incorrect metric may be the origin of flaws in dose–effect relationships for annoyance or sleep disturbance outcomes, which are mainly used for noise limit settings due to its importance at the population level.

This Special Issue was launched to promote a subject which is deserving of more attention: the study of new metrics, indicators or evaluation methods for noise exposure, and the relationship of noise with annoyance or other health effects, thus not relying only on an average noise exposure measure.

2. Summary of Published Papers

Most noise limits set by countries are expressed in absolute sound pressure levels, with only a few of them set with respect to the background noise level. This leads to the definition of sound emergence or, more generally, differential noise indicators. Many decades have passed since their introduction into legislation, when sound emergence was intended to limit the alterations of existing soundscapes which, in practice, raised several challenges both for the operators of noisy facilities, community, noise consultants, and authorities. Dutilleux et al. [16] established the concept of sound emergence as defined by international standardization in order to evaluate its relevance from different perspectives and to show that it was difficult to implement without bringing any benefit with respect to sound-pressure-level-based limit values on most occasions. The relevance of sound emergence was assessed from the point of view of perception and annoyance, in measurement practice and in development planning. Sound emergence seems to poorly predict audibility; the authors suggest considering the temporal and spectral characteristics of a specific source in the estimation of sound emergence, proposing source-specific choices of the metrics used for estimating total and residual sound pressure levels. Even if further research on the potential correlation between annoyance and sound emergence from specific sources is necessary, the best solution proposed is to combine an estimation of the audibility of specific sounds for the community and the estimation of a specific metric (Lspec).

Asensio et al. [17] investigated the scientific literature in research for the minimum set of indicators that would cover all physical dimensions of sound environments: energetic, spectral, and temporal dimensions, emergence, and source-related indicators. The extended set of indicators is intended to allow a more detailed analysis of the changes in noise environments related to confinement in the pandemic era and to a broader understanding the impact on sound environments of any policy achieved at the urban scale. In fact, noise indicators generally deal with sound environments considered as a whole, although they do not distinguish between the sound sources that compose it. Lockdowns imposed due to COVID-19 dramatically changed the noises emitted in terms of sound levels and by the different active sources. Reducing masking sounds made natural sounds perceptible again. Current indicators fail to represent this situation, although new source-orientated indicators would be able to quantify the presence of sources of interest. The authors conclude that the new physical indicators are associated with perceptual and health effects, but they will probably not be sufficient to represent the entire sound experience because sensitive and personal data or the connection between the sound environment and emotional evocations cannot be captured by physical indicators.

Differentiated studies for the singular type of sound sources were instead carried out by other authors published in this Special Issue, where most major sound sources have been investigated. From a point of view of the effects of noise on health, Petri et al. [18] evaluated the relationship of blood pressure and hypertension with noise produced by road, railway, airport, and leisure sources. In order to do so, noise measurements, blood pressure measurements, and a structured questionnaire were conducted on a significant set of patients. Noise exposure during the nights and diastolic blood pressure resulted correlated, with particular spikes for elders, moderately annoyed, noise-sensitive, without noise protection in the house and residents who usually did not close windows. In the investigated population, railway noise was the most impacting sound. The study also demonstrated an increase in the risk of hypertension in association with environmental noise.

Laboratory listening experiments were performed by Schäffer et al. [19], with the aim of searching for associations of annoyance and cognitive performance with the macro-temporal pattern of indoor exposure to noise emitted by road traffic. Relative quiet time and quiet time distribution were among the different metrics computed for the temporal pattern of the scenarios. Noise annoyance decreased with the increasing total duration of quiet periods, while shorter but regular breaks were less annoying than longer but irregular breaks. Different results were found for cognitive performance.

A lot of attention has been paid to interior noise in the present Special Issue. Li and Zheng [20] investigated passive noise control equipment as a means to control indoor environmental noise in contrast with the common applications of sound absorption materials and vibration isolation. According to the authors, active noise control has developed sufficiently to be effectively in avoiding the deficiencies of passive noise control, which are good mitigators for medium- and high-frequency noise but require a complex equipment and studies; generally, the low-frequency control effect is poor. In active noise control, loudspeakers are used to suppress noise in a specific area. Loudspeakers are omnidirectional; therefore, the sound pressure levels are reduced in the target area but are increased somewhere else. In their study, the authors calibrated a parametric array loudspeaker to achieve noise control in the target area in order to obtain a noise reduction topping 15.1 dB. The use of parametric array loudspeakers producing high-directivity sounds resulted to significantly reduce the noise interference to other adjacent areas while making the noise reduction areas more controllable.

Eventually, this solution would be beneficial even inside helicopter cabins, where Deacounu et al. [21] highlight the noise spectral components with measurements performed in different areas. High sound pressure levels were measured, with the urgency of reducing the exposure for passengers and crew. Main sources were identified in the transmission gear and the door area, with values during flight ranging from 97.2 dB(A) on the tail to 106.5 dB(A) under the transmission gear. Although the equivalent sound pressure levels vary spatially, the authors show that a much higher peak is registered at 2 kHz, and that this causes the most discomfort. This confirms once again how tonal sounds are more disturbing than others.

Railway noise and vibrations were investigated by Yan et al. [22], who performed measurements inside a metro train. Using A-weighted and linear sound pressure levels as metrics, and the FFT method for vibration measurements, the authors confirmed that interior sound levels increase sharply with the train's speed, while acceleration levels of the floor and sidewall are not apparent. Moreover, floor vibration contributed to the low-frequency noise components of the interior noise, and the characteristics of the vehicle dominate the frequency peaks of the acceleration levels of the floor and sidewall.

Leisure noise, together with the other major sources, have been investigated innovatively by Peplow et al. [23], using big data, i.e., citizens' tweets. The authors introduce a method, based on Python language, for estimating both the prevalence and location of noise annoyance. The open-access result produced would be also usable for the live monitoring of noise issues by means of tagging noise complaints.

Even port noise has been investigated in the Special Issue, with Schiavoni et al. [24] providing a review of recent findings in the scientific literature regarding port noise sources, an argument that has only quite recently been studied [25–28]. The database produced is expected to be useful to experts as inputs to the noise mapping phase, which is a fundamental step in the prevention of noise exposure. Recent decades have shown much focus from scientific community in this regard, with studies aimed at mitigating sound levels produced by almost all sources. Thus, quieter environments are expected. Unfortunately, Vukić et al. [29], by means of questionnaires on fishermen, showed that there are still sectors or environments where the impact of noise on health has not yet been accepted as a real danger and remains underestimated. Full social awareness can be reached only with proper education. In fact, the authors showed that when the test subjects underwent additional training, their practical knowledge and awareness of the noise impacts on health and society improved.

3. Conclusions

This Special Issue on the theme of the New Indicators for the Assessment and Prevention of Noise Nuisance has attracted the interest of authors from all over the world, publishing two reviews, two communications, as well as original research papers.

Progress has been made in the topic investigated, but it is still necessary to increase the awareness of the population, both in geographical terms and for workers in specific sectors, such as the marine industry.

It emerged that it is essential to carry out future studies that distinguish more between different sound sources with respect to their sound quality in terms of frequency, time pattern (fluctuation, emergence), and psychoacoustic indices, because a differential human reaction to sound sources is increasingly evident. More longitudinal studies are required. However, cross-sectional studies employing a more detailed soundscape description (including background) by competing sound indices are also useful to further the required knowledge to understand the human response in terms of the broad spectrum of potential adverse effects on health and quality of life.

Author Contributions: Conceptualization, L.F., P.L. and G.L.; methodology, L.F., P.L. and G.L.; writing—original draft preparation, L.F., P.L. and G.L.; writing—review and editing, L.F., P.L. and G.L. All authors have read and agreed to the published version of the manuscript.

Funding: This research received no external funding.

Conflicts of Interest: The authors declare no conflict of interest.

References

1. Themann, C.L.; Masterson, E.A. Occupational noise exposure: A review of its effects, epidemiology, and impact with recommendations for reducing its burden. *J. Acoust. Soc. Am.* **2019**, *146*, 3879–3905. [CrossRef] [PubMed]
2. Pienkowski, M. Loud Music and Leisure Noise Is a Common Cause of Chronic Hearing Loss, Tinnitus and Hyperacusis. *Int. J. Environ. Res. Public Health* **2021**, *18*, 4236. [CrossRef] [PubMed]
3. Teixeira, L.R.; Pega, F.; Dzhambov, A.M.; Bortkiewicz, A.; Silva, D.T.C.; de Andrade, C.A.F.; Gadzicka, E.; Hadkhale, K.; Iavicoli, S.; Martínez-Silveira, M.S.; et al. The effect of occupational exposure to noise on ischaemic heart disease, stroke and hypertension: A systematic review and meta-analysis from the WHO/ILO Joint Estimates of the Work-Related Burden of Disease and Injury. *Environ. Int.* **2021**, *154*, 106387. [CrossRef] [PubMed]
4. Basner, M.; Babisch, W.; Davis, A.; Brink, M.; Clark, C.; Janssen, S.; Stansfeld, S. Auditory and non-auditory effects of noise on health. *Lancet* **2014**, *383*, 1325–1332. [CrossRef]
5. Persson-Waye, K.; van Kempen, E.E.M.M. Non-auditory effects of noise: An overview of the state of the science of the 2017–2020 period. In Proceedings of the International Commission on Biological Effects of Noise, ICBEN, Stockholm, Sweden, 14–17 June 2021; pp. 1–35. Available online: http://www.icben.org/2021/ICBEN%202021%20Papers//full_paper_29017.pdf (accessed on 25 September 2022).
6. Guski, R.; Schreckenberg, D.; Schuemer, R. WHO environmental noise guidelines for the European region: A systematic review on environmental noise and annoyance. *Int. J. Environ. Res. Public Health* **2017**, *14*, 1539. [CrossRef]
7. Basner, M.; McGuire, S. WHO environmental noise guidelines for the european region: A systematic review on environmental noise and effects on sleep. *Int. J. Environ. Res. Public Health* **2018**, *14*, 519. [CrossRef]
8. Thompson, R.; Smith, R.B.; Bou Karim, Y.; Shen, C.; Drummond, K.; Teng, C.; Toledano, M.B. Noise pollution and human cognition: An updated systematic review and meta-analysis of recent evidence. *Environ. Int.* **2022**, *158*, 106905. [CrossRef]
9. Schubert, M.; Hegewald, J.; Freiberg, A.; Schubert, M.; Hegewald, J.; Freiberg, A.; Starke, K.R.; Augustin, F.; Riedel-Heller, S.G.; Zeeb, H.; et al. Behavioral and Emotional Disorders and Transportation Noise among Children and Adolescents: A Systematic Review and Meta-Analysis. *Int. J. Environ. Res. Public Health* **2019**, *16*, 3336. [CrossRef]
10. Dzhambov, A.M.; Lercher, P. Road traffic noise exposure and depression/anxiety: An updated systematic review and meta-analysis. *Int. J. Environ. Res. Public Health* **2019**, *16*, 4134. [CrossRef]
11. Daiber, A.; Kröller-Schön, S.; Frenis, K.; Oelze, M.; Kalinovic, S.; Vujacic-Mirski, K.; Kuntic, M.; Bayo Jimenez, M.T.; Helmstädter, J.; Steven, S.; et al. Environmental noise induces the release of stress hormones and inflammatory signaling molecules leading to oxidative stress and vascular dysfunction-Signatures of the internal exposome. *Biofactors* **2019**, *45*, 495–506. [CrossRef]
12. van Kempen, E.; Casas, M.; Pershagen, G.; Foraster, M. WHO environmental noise guidelines for the European region: A systematic review on environmental noise and cardiovascular and metabolic effects: A summary. *Int. J. Environ. Res. Public Health* **2018**, *22*, 379. [CrossRef] [PubMed]
13. Lim, S.S.; Vos, T.; Flaxman, A.D.; Danaei, G.; Shibuya, K.; Adair-Rohani, H.; Amann, M.; Anderson, H.R.; Andrews, K.G.; Aryee, M.; et al. A Comparative Risk Assessment of Burden of Disease and Injury Attributable to 67 Risk Factors and 21 Risk Factor Clusters in 21 Regions, 1990–2010: A Systematic Analysis for the Global Burden of Disease Study 2010. *Lancet* **2012**, *380*, 2224–2260. [CrossRef]
14. Badino, A.; Borelli, D.; Gaggero, T.; Rizzuto, E.; Schenone, C. Airborne noise emissions from ships: Experimental characterization of the source and propagation over land. *Appl. Acoust.* **2016**, *104*, 158–171. [CrossRef]
15. Fredianelli, L.; Gallo, P.; Licitra, G.; Carpita, S. Analytical assessment of wind turbine noise impact at receiver by means of residual noise determination without the wind farm shutdown. *Noise Control. Eng. J.* **2017**, *65*, 417–433. [CrossRef]
16. Dutilleux, G.; Gjestland, T.; Licitra, G. Challenges of the use of sound emergence for setting legal noise limits. *Int. J. Environ. Res. Public Health* **2019**, *16*, 4517. [CrossRef]
17. Asensio, C.; Aumond, P.; Can, A.; Gascó, L.; Lercher, P.; Wunderli, J.M.; Licitra, G. A taxonomy proposal for the assessment of the changes in soundscape resulting from the COVID-19 lockdown. *Int. J. Environ. Res. Public Health* **2020**, *17*, 4205. [CrossRef]

18. Petri, D.; Licitra, G.; Vigotti, M.A.; Fredianelli, L. Effects of exposure to road, railway, airport and recreational noise on blood pressure and hypertension. *Int. J. Environ. Res. Public Health* **2021**, *18*, 9145. [CrossRef]
19. Schäffer, B.; Taghipour, A.; Wunderli, J.M.; Brink, M.; Bartha, L.; Schlittmeier, S.J. Does the Macro-Temporal Pattern of Road Traffic Noise Affect Noise Annoyance and Cognitive Performance? *Int. J. Environ. Res. Public Health* **2022**, *19*, 4255. [CrossRef]
20. Li, Y.; Zheng, W. A Noise Control Method Using Adaptive Adjustable Parametric Array Loudspeaker to Eliminate Environmental Noise in Real Time. *Int. J. Environ. Res. Public Health* **2021**, *19*, 269. [CrossRef]
21. Deaconu, M.; Cican, G.; Toma, A.C.; Drăgășanu, L.I. Helicopter Inside Cabin Acoustic Evaluation: A Case Study—IAR PUMA 330. *Int. J. Environ. Res. Public Health* **2021**, *18*, 9716. [CrossRef]
22. Yan, L.; Chen, Z.; Zou, Y.; He, X.; Cai, C.; Yu, K.; Zhu, X. Field study of the interior noise and vibration of a metro vehicle running on a viaduct: A case study in Guangzhou. *Int. J. Environ. Res. Public Health* **2020**, *17*, 2807. [CrossRef] [PubMed]
23. Peplow, A.; Thomas, J.; AlShehhi, A. Noise annoyance in the UAE: A Twitter case study via a data-mining approach. *Int. J. Environ. Res. Public Health* **2021**, *18*, 2198. [CrossRef] [PubMed]
24. Schiavoni, S.; D'Alessandro, F.; Borelli, D.; Fredianelli, L.; Gaggero, T.; Schenone, C.; Baldinelli, G. Airborne Sound Power Levels and Spectra of Noise Sources in Port Areas. *Int. J. Environ. Res. Public Health* **2022**, *19*, 10996. [CrossRef] [PubMed]
25. Fredianelli, L.; Bolognese, M.; Fidecaro, F.; Licitra, G. Classification of noise sources for port area noise mapping. *Environments* **2021**, *8*, 12. [CrossRef]
26. Borelli, D.; Gaggero, T.; Rizzuto, E.; Schenone, C. Holistic control of ship noise emissions. *Noise Mapp.* **2016**, *3*, 107–119. [CrossRef]
27. Fredianelli, L.; Gaggero, T.; Bolognese, M.; Borelli, D.; Fidecaro, F.; Schenone, C.; Licitra, G. Source characterization guidelines for noise mapping of port areas. *Heliyon* **2022**, *8*, e09021. [CrossRef]
28. Nastasi, M.; Fredianelli, L.; Bernardini, M.; Teti, L.; Fidecaro, F.; Licitra, G. Parameters affecting noise emitted by ships moving in port areas. *Sustainability* **2020**, *12*, 8742. [CrossRef]
29. Vukić, L.; Mihanović, V.; Fredianelli, L.; Plazibat, V. Seafarers' Perception and attitudes towards noise emission on board ships. *Int. J. Environ. Res. Public Health* **2021**, *18*, 6671. [CrossRef]

International Journal of
Environmental Research and Public Health

Review

Challenges of the Use of Sound Emergence for Setting Legal Noise Limits

Guillaume Dutilleux [1,*], Truls Gjestland [2] and Gaetano Licitra [3,4]

[1] Acoustics Research Center, Department of Electronic Systems, NTNU, 1,7491 Trondheim, Norway
[2] Acoustics Research Center, SINTEF Digital, 1,7491 Trondheim, Norway; truls.gjestland@sintef.no
[3] Physics Department, University of Pisa, Largo Pontecorvo 3, 56127 Pisa (PI), Italy; G.licitra@arpat.toscana.it
[4] Environmental Protection Agency of Tuscany Region, Via Vittorio Veneto 27, 56127 Pisa (Pi), Italy
* Correspondence: guillaume.dutilleux@ntnu.no; Tel.: +47-73-55-92-32

Received: 21 October 2019; Accepted: 10 November 2019; Published: 15 November 2019

Abstract: In the vast majority of legislation on environmental noise, the metric used for expressing limit values is based on sound pressure levels. But some countries have introduced sound emergence limit values where the compliance of a noise-generating activity is defined as a maximum allowable difference between the sound pressure level with and without the regulated activity operating. This paper investigates the foundations and the merits of this kind of differential noise limit values. Our review of literature indicates that there is very little evidence supporting the use of differential noise limits over absolute ones. Moreover, while sound emergence limits seem to originate from consideration about audibility of the regulated noise source, they appear to give little insight into what is audible and what is not. Furthermore, both the definition and the practical measurement of sound emergence raise several challenges that compromise reproducibility. In addition, first, the reference to background noise makes it very difficult first to ascertain the conformity of noisy installations in the long run, second to effectively protect the community from excessive noise and third to evaluate conformity on the basis of simulations. When switching to another metric is not an option the paper makes recommendations toward a more reliable use of sound emergence.

Keywords: sound emergence; legislation; annoyance; measurement; prediction; uncertainty; audibility; signal-to-noise ratio

---

1. Introduction

A great many of countries in the world have evolved regulations against environmental noise, including noise from transportation, industry and community noise [1,2]. Most of these countries have chosen to rely on limit values expressed in "absolute" sound pressure levels. A few of them, however, prefer to express limit values with respect to the background sound pressure level, at least regarding community noise or industry noise. Relating the total noise or the source-attributable noise to background noise leads to the concepts of *sound emergence* or more broadly speaking what can be coined *differential noise indicators*. Sound emergence is indeed an instance of the class of differential noise indicators. Especially at the time when legislations against noise were developed, there was little evidence to justify one type of metric against another. Sound emergence was clearly an option among many others like the continuous equivalent sound pressure level, the maximum level based on the time-weighted sound pressure level or sound exposure. Initially, sound emergence may seem quite relevant with the intention to limit the alterations to existing soundscapes in mind. But in practice the implementation of sound emergence is not straightforward.

The purpose of this paper is to discuss in further detail the concept of sound emergence as defined by international standardization, to evaluate its relevance from different perspectives and to show that it raises several challenges both for the operators of noisy facilities, for the community, for noise consultants and authorities, without bringing any benefit with respect to sound-pressure-level-based limit values in most occasions. Only an articulate legislation can help overcome some of the pitfalls of sound emergence evaluation.

This paper is organized as follows. Section 2 provides the necessary definitions of sound emergence and its components. Section 3 reviews the presence of the concept of sound emergence in the legislation in a large subset of developed countries, and makes the distinction with related but clearly different indicators. Section 4 evaluates the relevance of sound emergence with respect to human perception and annoyance. Section 5 deals with the underlying challenges of the sound emergence when it comes to implementing this metric in measurement standards. The impracticality of the use of sound emergence in development planning is covered in Section 6. Section 7 brings a few recommendations.

2. Definitions

Sound emergence has been present in ISO 1996 since the first release of the standard [3]. It is defined as follows in the current ISO 1996-1 [4]:

Definition 1. sound emergence [[4], §3.4.7] *increase in the total sound in a given situation that results from the introduction of some specific sound*

where

Definition 2. total sound [[4], §3.4.1] *totally encompassing sound in a given situation at a given time, usually composed of sound from many sources near and far*

and

Definition 3. specific sound [[4], §3.4.2] *component of the total sound that can be specifically identified and which is associated with a specific source.*

The specific source is typically the source the noise impact of which is to be evaluated. For the sake of the discussion it is necessary to define the residual sound:

Definition 4. residual sound [[4] §3.4.3] *total sound remaining at a given position in a given situation when the specific sounds under consideration are suppressed.*

These definitions are identical to the ones in the previous release of ISO 1996-1 [ISO1996-1:2003].

In the literature residual sound in the ISO 1996-1 sense is often referred to as the *ambient sound* (see for instance [5]). This can be confusing because *ambient sound* is also used as a synonym of the above-defined *total sound* in other documents [6]. Therefore, *ambient sound* is not used in the remainder of this paper.

It is convenient to identify sound emergence as e to reduce the risk of confusion with sound exposure "E" [7]. In practice e is defined as the subtraction of the residual sound pressure level L_{res} from the total sound pressure level L_{tot}.

$$e = L_{tot} - L_{res} \tag{1}$$

e is expressed in decibels. The metrics used for L_{res} and L_{tot} will be discussed later on. To complete the notations, L_{spec} stands for the specific sound level in the following.

The concepts of total sound, specific sound and residual sound can be conveniently illustrated as in Figure 1. Depending on the context, a source can be either considered as the part of the residual sound or as the specific source.

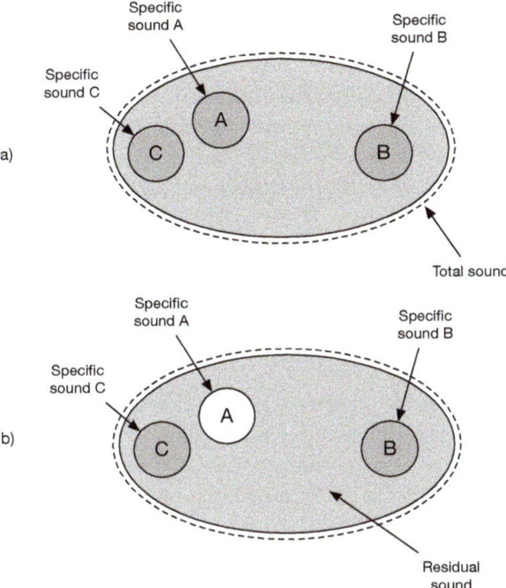

Figure 1. Total, specific and residual sound. 3 sources A, B and C are identified but other non identified sources combine into the total sound (**a**). With respect to A, the residual sound is observed when the specific sound A is absent (**b**), everything else being equal.

Although sound emergence is defined in ISO 1996-1 as shown above, the definition is the only occurrence of the concept in the ISO 1996 series and the standard neither does elaborate on how sound emergence can be obtained nor clarify its meaning. The details of implementation are left to national standards. There are, however, two possible interpretations of the *increase* referred to in the definition of sound emergence. This is discussed further in Section 5.4. Indeed, the ISO 1996 series focuses on he determination of sound pressure levels.

3. Sound Emergence in Current Official Documents

3.1. Sound Emergence Stricto Sensu

World Bank appears to use the concept of sound emergence without naming it both in general guidelines [8] and specific ones [9] by specifying a maximum increase in *"background levels of 3 dB at the nearest receptor location"*. This guideline is always combined with limit values expressed in L_{eq} and applies beyond the property boundaries of the noisy facilities.

While noise limits in sound pressure level are used in France for transportation infrastructure noise, in the case of industry and community noise, this country specifies limit values with respect to the difference between the total sound pressure level and the background sound pressure level [2,10]. To our knowledge, the first occurrence of such noise differential noise limits is found in a legal text [11]. This difference between the total and the background sound pressure level is called *émergence*. In the current legislation it should not exceed 5 dB(A) in day time and 3 dB(A) in night time when the total sound pressure level is higher than 45 dB(A). When the sound pressure level is between 35 dB(A) and 45 dB(A) the limit values become 6 dB(A) and 4 dB(A) respectively. The French legislation relies on NF S 31-010 standard for the measurement of emergence in this case [6]. At the time of writing, this standard is under revision. In the dedicated legislation on wind turbine noise, France sets emergence-based limits at $5 + k$ dB(A) in day time (7:00–22:00) and $3 + k$ dB(A) in night time (22:00–7:00) when the total sound level exceeds 35 dB(A) [12] where $k = 0$ dB when the noise from the park is *apparent* for more than 8 hours over 24 hours. For shorter durations k ranges from 1 to 3 dB. Here, the French legislation refers to pr S 31-114 draft standard on wind turbine noise assessment for the practical aspects [13] where sound emergence is based on L_{A50}. The French management of community noise is also based on emergence in dB(A) or in octave bands [14]. In dB(A) a similar approach is used as for industry noise but k can take values from 1 to 6 when the cumulated duration of occurrence of the noise to be regulated decreases from below 8 hours to 1 minute. Spectral emergence should not be confused with tonality [15]. The former is defined within an octave band. The limit values for spectral emergence are 7 dB for the octaves 125 to 250 Hz and 5 dB for the octaves 500 to 4000 Hz. The French legislation on places where sound reinforcement is used sets to 3 dB the maximum emergence in the octaves 125 to 4000 Hz [16]. L_{A50} is often specified or suggested for the estimation of the background sound pressure level but other indicators are allowed in the case of community noise [6].

Since 1991 [17], Italy defines so-called *differential noise limits* that correspond to sound emergence [18,19]. This criterion applies to the noise from industrial facilities only, including wind turbines [20] outside areas that are classified as industrial ones. The thresholds are set to 3 dB in the night time and 5 dB in the day time, like in France. In Italy, however, sound emergence is only measured indoors, and it is always combined with immission and emission noise limits [18]. With windows open sound emergence applies only if L_{tot} is larger than 50 dB(A) in the day time and 40 dB(A) in the night time. With windows closed, these thresholds fall down respectively to 35 dB(A) and 25 dB(A). In Italy, sound emergence is strongly oriented to the protection of the receiver since the worst case between windows open and windows closed is used for evaluating conformity. The motivation for the introduction of sound emergence into the Italian legislation is annoyance. The distance to the source does not matter since the measurement is carried out at the receiver. The only facilities for which the differential noise limit does not apply are plants built before 1996 that operate uninterruptedly, acknowledging the impossibility to evaluate the difference between L_{tot} and L_{res}. However, as soon as modifications occur in the existing installations, the differential noise limit enters into force [21].

3.2. Other Ways to Refer to L_{res} When Setting Noise Limits

The is some confusion around the concept of sound emergence. Some authors indicate that Australia and the United Kingdom use sound emergence in the ISO 1996-1 sense [22] in the case of wind turbine noise. Another paper seems to concur by stating that the UK and Australia have the same approach since they enforce a comparison of L_{Aeq} with the background sound level when assessing conformity [23]. In either case what is overlooked is the distinction between L_{spec} and L_{tot}. Only the latter is consistent with sound emergence. Furthermore, legislation may vary within a country, as it is the case for Australia. Among the regulations publically available in English, German and in any of the Scandinavian languages, we could not find any document where the comparison to background noise is made on L_{tot}.

In Ireland, the recommended approach is to use rating sound pressure levels to set maximum allowable contributions from licensed sites [24,25]. This limit value can occasionally depend on the background noise level. In the case of wind turbines the proposed limits for licensed sites rely on the principle that turbine noise should be controlled with reference to absolute limits when background is low, or relative to background noise itself as background noise increases with wind speed, whichever is greater. In practice, this principle is interpreted so that turbine-attributable noise should be limited to either a certain L_{A90} or to 5 dB above the background noise [25].

The United Kingdom appears to use differential noise limits for rating and assessing industrial and commercial sound [26]. The same applies to wind farms [27]. However, the assessment criterion deviates significantly from sound emergence. The main difference is that the aim is not to obtain the total sound pressure level but the *source-attributable* noise or L_{spec} from L_{tot} (called ambient sound level in [26]) via a classical background noise correction. Moreover, by comparison to the French and Italian regulations, the general practice in the UK, at least in the case of wind farm projects is to estimate L_{res} well ahead of the measurement of L_{spec}. In the case of industrial and commercial sound, a value of the difference between the rating sound pressure level of the specific sound source and the background sound pressure level that is equal to or higher than +10 dB (resp. +5 dB) is deemed "likely to be an indication of a significant adverse (resp. of an adverse) impact". These guidelines acknowledge explicitly that the impact is context-dependent. The noise indicators specified in [26] are L_{Aeq} for L_{tot}, L_{spec}, L_{res} and L_{AF90} for background noise. The background noise is not identical to the residual sound (Cf Section 2) since background noise is estimated when the source under investigation is operating whereas L_{spec} assumes that the source under investigation is turned off. Regarding the assessment of wind farm noise, the indicator to be used is $L_{A90,10min}$ [27]. When the difference between L_{tot} and background sound level is lower than 3 dB at the receiver, the measurement of L_{spec} is ill-conditioned. It is then recommended to choose measurement points closer to the source where the signal-to-noise ratio (SNR) is high enough so that the sound power of the source can be estimated. L_{spec} at the receiver [26] can then be obtained by simulation. The differential threshold for wind farms noise is set to 5 dB [27].

In Australia, noise legislation varies from one state or territory to another. A survey of the existing legislation was provided in [28]. While this paper was published in 2003, the situation remains essentially the same, even though most of the state-specific guidelines have been revised. The reference to background sound in noise limits is well represented across the country, since at least New South Wales (NSW), South Australia (SA) and Victoria use the so-called "background-plus rule" in their guidelines [29–31], where the increment to background noise is typically 5 dB. Moreover, Tasmania states that the compliance of a new development must be assessed by comparison of its L_{spec} with the background noise [32], although to our knowledge the trigger value is not clearly stated. For night time, the 5 dB increment is replaced by 0 dB for local government issues in NSW [29]. In addition, in the case of noise from industry, a so-called "rating background noise level" is substituted for the measured background noise level in order to set a conventional lower limit to background sound pressure level [33]. The background sound pressure level is also used at least in NSW, Queensland, SA and Tasmania to correct for extraneous noise in the assessment of L_{spec}. But at least three of the five Australian states or territories who set noise limits with reference to background sound pressure level, also set absolute noise limits in parallel, like NSW, Queensland and SA.

4. Emergence, Annoyance and Perception

4.1. Sound Emergence Is a Second Order Descriptor of Annoyance

It is well known that noise annoyance is only partly determined by acoustic factors [5,34]. One can expect, however, that the metrics used for setting noise limit values be connected with health effects in the broad sense, and annoyance in particular. It appears that very little research was done on the merits of sound emergence from a public health perspective. This was already pointed out more than two decades ago [35] and could not be contradicted by our own investigations. We found only two papers dealing explicitly with annoyance and sound emergence. The first one focuses on noise sources relating to electric power generation [36]. The second one addresses annoyance from impulsive sounds [37]. Another paper quoted by [35] evaluated the so-called *salience*, defined as $L_{Aeq,10ms} - L_{Aeq,1s}$, among different candidate ratings against a subjective one for impulsive noise [38]. In this context, salience can be considered as an instance of sound emergence. Both references found about impulsive sounds conclude that criteria not referring to background sound pressure levels perform better than sound emergence. Moreover, to our knowledge there is no published dose-response curve based on sound emergence.

On the contrary, a wide meta-analysis on 136 surveys concluded among other things that "noise annoyance is not affected to an important extent by residual sound levels" [5,39]. The surveys used are screened from a very large sample of field surveys. The selection was carried out on five criteria: (1) is the effect of the residual as strong as the one of a 3 dB(A) increase in the specific sound pressure level, (2) is the residual sound causing a 5% increase in the number of annoyed, (3) does the residual explain 1% of the variance in annoyance, (4) is the effect of the residual sound statistically significant and (5) is there any verbal association between residual sound and annoyance. However the surveys analyzed dealt mostly with aircraft noise and road traffic noise as the source of specific sound. Nonetheless transportation noise encompasses a wide variety of sounds, both steady and unsteady, with or without tonalities and the various levels of ownership among the respondents with respect to the source of specific sound. Another limitation of this analysis is that the lower level end of the range of sound pressure levels is somewhat underrepresented in the surveys analyzed.

4.2. Emergence Is only a Proxy for Signal-to-Noise Ratio SNR When SNR Is High

Under the incoherent summation hypothesis sound emergence can be related to the signal-to-noise ratio SNR in dB

$$\text{SNR} = 10 \log_{10} \frac{\tilde{p}^2_{spec}}{\tilde{p}^2_{res}} \tag{2}$$

where $\tilde{p}$ is an estimator of the root-mean-square value of the acoustic pressure, subscript *spec* refers to the specific sound and *res* to the residual sound. From Equation (1) one can write

$$e = 10 \log_{10} \frac{\tilde{p}^2_{tot}}{\tilde{p}^2_{res}} = 10 \log_{10} \frac{\tilde{p}^2_{res} + \tilde{p}^2_{spec}}{\tilde{p}^2_{res}} \tag{3}$$

$$= 10 \log_{10} \left(1 + 10^{\frac{\text{SNR}}{10}} \right). \tag{4}$$

As illustrated in Figure 2, e is a good proxy for SNR provided that SNR is higher than about 10 dB. But the range of SNR$\leq$ 0 dB is compressed into a very narrow range of sound emergence since this SNR range maps onto $0 \leq e < 3$ dB where $e = 3$ dB corresponds to SNR $= 0$ dB.

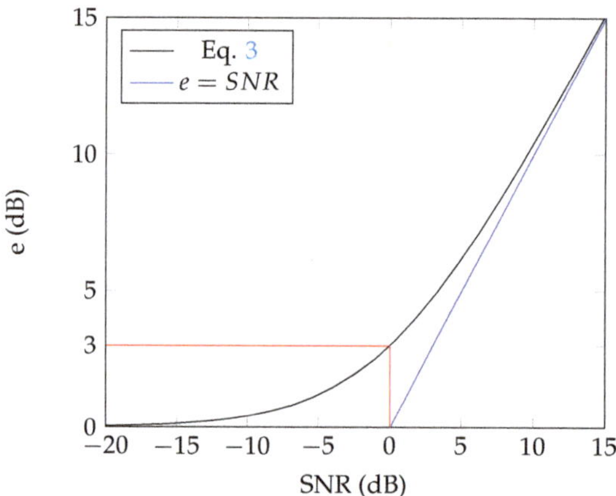

Figure 2. Relationship between signal-to-noise ratio (SNR) and e.

4.3. Sound Emergence Does Not Reflect Detectability

The origins of sound emergence are not clearly established. However, if one takes into account (1) that the use of this indicator is limited to lower sound pressure levels, (2) that national standards implementing sound emergence refer explicitly to detectability and (3) that sound emergence is defined by reference to the residual sound pressure level, a possible hypothesis is that sound emergence was introduced as a proxy for the detectability of a specific source in the soundscape. It is well known that a sound that reaches a person's consciousness and is clearly identified can be extremely annoying, even though is sound pressure level is low [40]. The connection between SNR and e can help assess the capacity of emergence to state whether a specific sound can be heard or not.

Detectability of low-levels has been investigated in [41] for a variety of acoutic simuli ranging from 38 to 70 dB(A) combined with different types of masking sounds. This research shows a strong correlation between subjective judgement and a measure of detectability that is proportional to SNR. Moreover the same authors provide evidence that detectability is not limited to positive values of SNR [42]. For the specific case of wind turbines, this is further documented in more recent research where wind turbine sounds are detected at SNR $= -8$ dB [43]. In addition, [44] estimates the detection limit of wind turbine in presence of highway noise to -23 dB(A). With such thresholds of detectability, the current typical limit values at $e = 3$ dB for night time fail to ascertain that the specific sound will not be detected by a major part of the population living in the neighborhood of the wind farm if this limit is satisfied.

While it has been illustrated in Figure 2 that e is very close to SNR when SNR is strongly positive, it is not the case when SNR < 0. As a conclusion, e does not reflect audibility in the range of SNR where a specific source is likely to be detected and it does not help decide whether a specific sound can be heard or not.

4.4. The Relationship between Annoyance and the Strength of Sound Emergence

Let us assume now that sound emergence is intended to be a proxy for annoyance. There is strong evidence that different noise sources are not equally annoying at the same equivalent sound pressure level [15,45]. This is taken into account in measurement standards that consider different penalties to account for impulsiveness, low frequency content or tonality [15]. In the existing regulations that use sound emergence, however, the threshold is not source-dependent but common to the wide categories of so-called industrial noise or community noise. There is however evidence that not all industrial noise sources are equally annoying at the same sound emergence level [36]. The research documents that even when $e > 5$ some sources are not deemed annoying. This is attributed to the wide spectrum of the specific sound that shapes the spectrum of the residual sound.

5. Measuring Sound Emergence

5.1. Measurement Uncertainty

Since it is defined as a difference of sound pressure levels, the uncertainty budget of sound emergence is less favorable than the one of an individual sound pressure level. Everything else being equal, in the general case where there is no correlation between L_{spec} and L_{res}, the total uncertainty attached to sound emergence is the geometric mean of the uncertainties of the two sound pressure levels [46]. Even with the best measurement equipment available on the market and the most favorable measurement conditions the metrological uncertainty is not likely to be lower than 0.7 dB when a class I sound level meter is used [47]. Assuming a 95% confidence interval and an unilateral interval, the extended uncertainty is 1.15 dB in this case. The presence of tonal components in the signal may lead to significantly larger instrument-related uncertainties [48]. A more realistic estimate should include representativity and reproducibility components. If these elements are taken into account, it is very unlikely that the extended uncertainty will be lower than 1.5–2 dB. Moreover if one takes into account that legal threshold values can be as low as 3 dB, establishing the compliance of a noise source will be problematic a soon as uncertainties are taken into account.

5.2. Specific Sound in Practice

In practice, deciding whether the specific sound is present or not is a matter of perspective. At least two points of view can be distinguished: from the source or from the receiver.

From the perspective of the source, the specific sound is present in the total sound when the source of specific sound is radiating sound. The advantage of this approach is that documenting source operation is quite straightforward and reliable by using acoustical means, provided that one has access to the vicinity of the source. Non-acoustical means can also be used, for instance in the case of a factory where the operating hours and the manufacturing process are well known. But the fact that the source of specific sound is active does not imply that it is perceived by the receiver.

In the perspective of the receiver the specific sound is present in the total sound if the specific sound is audible. In general, audibility can not be documented by a sound level meter. It will rather require the presence of a human operator who will be able to detect the specific sound in the total sound. By the current state of technology this task is not easily automated, although blind source separation is a quite active area of research [49]. Some authors have developed a dedicated method for the separation of wind turbine noise from background noise in order to assess the conformity of a wind park with respect to the Italian legislation [20]. The procedure developed makes, however, strong assumptions about background noise, namely that residual sound only depends on wind and that wind on the ground is not correlated to wind at hub height. This hypothesis has been reported to work in rural areas [50] but it will not be satisfied everywhere. Furthermore the method may not generalize easily to other community or industrial noise sources.

The receiver perspective seems more relevant when it comes to evaluating community response. However it makes the estimation of specific sounds operator-dependent. With standardization of field noise measurement procedures in mind, this is problematic because it would compromise measurement reproducibility. First, although the average capacity of a healthy human hearing is well documented [51], the dispersion around the average is still to be assessed [52]. Second, hearing loss is a pathology that can go unnoticed for a long time especially when hearing loss does not affect speech comprehension [53]. Moreover hearing tests are not a routine practice of occupational health check-ups in many countries. Third, when listening for specific sound, higher level capacities of the human operator may interfere like attention, concentration and the knowledge of the variability of the source of specific sound.

The receiver perspective also sets limits to the duration of the interval of observation. In the absence of machine-based estimation of specific sound, this task can only be performed over short term measurements, i.e., not beyond a few hours of listening while arguably the compliance of a noisy facility should in general be assessed over a longer time frame for the sake of representativeness. As the variability of the source under investigation and the range between source and receiver increases, so does the uncertainty of the estimation of specific sound.

5.3. Residual Sound in Practice

While residual sound is easily told from the total sound in a picture like the one of Figure 1, in reality it is not always straightforward to estimate L_{res}, because what is captured by the microphone is the total sound. The source is generally not under the control of the operator carrying the noise impact assessment.

Moreover, in several cases the source operates permanently with a quite stable sound emission. Stopping such a source can be either costly or simply not an option. The power network provides several examples where the estimation of residual sound is challenging. First, turning off a wind farm is possible but results in large losses of revenue for the operator. Second, stopping a large installation like a nuclear power plant is a very long process that raises the issue of the redundancy of the power network so that the loss of production can be compensated by another plant. At the other end of the power networks evaluating the sound emergence of power transformer means a power outage for hundreds of people or strategic services like a hospital.

As already mentioned for the specific sound, blind source separation is not currently available off-the-shelf. In addition, it will be detailed later, modeling residual sound as a substitute for measurements is a difficult task with many unknowns and large uncertainties that are not compatible with low emergence-based limit values. To solve this issue, a common approach is to estimate the residual sound pressure level at another location far enough from the source of specific sound that cannot be stopped. This is of course problematic for both reproducibility and representativeness. Alternatively, it may be possible to benefit from maintenance phases to measure the residual sound pressure level at the right location but then facing the risk of non-contemporaneousness between the estimation of the specific source and the one of the residual sound. The puzzle of the unstoppable source has no perfect solution and all noise limits that refer to L_{res} face this issue, not only sound emergence.

5.4. Incoherent Source Assumption

The meaning of "increase" in the definition of sound emergence can also be a matter of interpretation. Currently, "increase" can be understood in the algebraic sense, so that e can take negative values. Under the assumption of coherent sources, assuming that the residual sound contains sound from a sound source A it is theoretically possible to observe a local decrease in sound pressure level when a sound source B is turned on, especially if A and B produce tonal noise at the same frequency. This may happen in practice for instance in the case of factories that combine several identical units that generate low frequencies.

Destructive interferences, however, are not expected for most of real world noise sources at the usual receiver distances. Indeed, the implicit assumption behind sound emergence is the one of incoherent summation of contributions from residual sound and specific sound so that $e \geq 0$ dB.

5.5. Variety of Metrics

As mentioned, the ISO 1996-1 standard does not go into details regarding the practical evaluation of sound emergence and especially what metric to use for the quantification of the sound pressure level of specific sound. This aspect is at least partially addressed in a national standard [6]. This French standard leaves the measurement operator free to choose a suitable measure between equivalent sound pressure levels $L_{eq,T}$, where time constant T is not specified and fractile sound pressure, levels L_X, where the threshold is not specified either, derived from L_{eq} time series. The wide freedom left regarding the specification of the metric used for the residual sound and for the total sound may reflect that there is no consensus on suitable metrics for the evaluation of different environmental sound mixtures. The consequence is that different noise consultants may decide to choose different metrics for the same source which can not but compromise the reproducibility of emergence measurements.

6. Emergence and Development Planning

Since emergence is defined with respect to residual sound, setting sound-emergence-based noise limits is problematic for all the stakeholders because they are both difficult to predict and difficult to ascertain in the long run.

6.1. Seen from the Source

First, emergence is problematic for the owner or the operator of a noisy facility, as emergence-based noise limits offer little assurance about the long-term compliance of the facility to the noise limits. One can assume that the operator has a good command on the noise emissions of the facility and that these emissions can be kept at the same sound pressure level as specified or observed during the environmental impact assessment.

But, conversely, the operator has little control over variations in background noise. For instance a plant A could be located close to another noisy plant B. In the environmental impact study for plant A the noise from plant B contributes to the definition of the residual sound in the neighborhood of plants A and B. If plant B goes bankrupt or is relocated, the decrease in residual sound level is to be expected and the compliance of plant A with respect to noise is at stake. Other changes in the environment could have similar consequences like the construction of a building or a barrier between plant B and the community. This would obviously reduce the contribution of plant B to the residual sound level and may compromise the conformity of plant A.

Since it is very difficult to predict sound emergence using software simulation tools, emergence will be measured. But the duration of the measurement campaign will be constrained by financial and practical considerations. Therefore, the measurements will typically not be carried out over more than a few days. This prevents the documentation of seasonal variations. If the environmental impact assessment is done well into the vegetation period, then background noise may be strongly influenced by the interaction of wind with foliage while this component will be more or less absent outside the vegetation period in the case of deciduous trees. But vegetation is not the only parameter subject to seasonal variations inducing seasonal variations in residual sound levels. One can also mention seasonal winds, snow cover and seasonal human activities.

6.2. The Community Perspective

Second, emergence is problematic for the community because the reference to residual sound in the noise limit sets a shifting baseline [54]. In other words, it offers little protection, if any, against higher noise levels. Let us imagine a pristine rural environment without any noisy activity. Plant A may be allowed to operate continuously after the environmental impact assessment because the immission level in the community is not higher than $L_{res} + k$ decibels. If a few years later another factory C wants to operate in the very same area, the reference residual sound level for factory C will then be $L_{res} + k$, everything else remaining equal, and the consequence of the continuous operation of factory C may lead to sound levels as high as $L_{res} + 2k$. As new economic developments appear in the area the sound can still increase due to the shifting baseline of the residual sound level. Unless cumulative noise limits are introduced like the one defined in [25] or in the Italian legislation [21], the limit will only be set by shortage of available land.

6.3. Predicting Emergence

Third, emergence is problematic because it is very difficult to predict with good confidence. Again, the main issue is the reference to residual sound. Residual sound is a priori a mixture of a wide variety of sources. This raises several challenges. The first is to identify the sources. They may include streams, foliage noise from trees and shrubs, birds during the dawn chorus, elongated structures singing in the wind, diverse appliances and equipments present in the environment—like heat pumps—road and rail traffic and industrial noise. This implies a survey over a wide area.

Provided that this survey be successful, the physical modeling of the emission, the spatial distribution, the duty pattern or cycle of these multiple sources is not an obvious task and requires a lot of effort. Some of the sources listed may seem of minor importance and it is certainly possible to perform a ranking and focus on the most contributing sources but in rural settings, like in the case of wind energy developments, it is likely that the simulation of the residual sound level would require the consideration of sources, the modeling of which is not well established. Moreover, at the time of writing there is no such thing as a reliable macroscopic model for L_{res}. The general spectral trends are well known [55] but the calibration is problematic and the evaluation of emergence must always be carried at a specific place with specific sources.

Furthermore, an additional difficulty is brought by the fact that L_{res} may be as low as 30–35 dB. It is well known that the current engineering-level noise prediction methods are not designed for the simulation of such low levels. All this is in stark contrast with the general approach in noise impact studies where limit values are expressed as sound levels. Simulations are a routine operation and if the regulations set limits on the contribution of the source under study, the modeling can focus on the infrastructure/plant/source under investigation and forget the other sources which is a much more feasible task. This allows to investigate yearly-averaged values by considering the seasonal variations of the emissions and of the propagation medium and also to consider long term trends. All of this is beneficial to the stability of the compliance of noisy installations.

7. Conclusions and Recommendations

We reviewed the use of sound emergence limit values in legislation and other official documents. Noise limit values referring to background sound appear to be in use in a small number of legal texts throughout the world. The background sound level can be used to define the initial state of the soundscape in a place before setting the maximum allowable contribution for the specific sound source to be regulated or for setting the allowable sound pressure level of the total sound. Some countries, however, prefer to express the noise limit in real time by a direct reference to the difference between total sound pressure level and residual sound pressure level. Occasionally the allowed difference can be as low as 3 dB(A) and is supposed to be obtained from measurements.

In this paper the relevance of sound emergence was assessed from the point of view of perception and annoyance, the one of measurement practice and the one development planning. The literature indicates that there is little evidence, if any, that sound emergence is a better rating than sound pressure level because the background sound is a second order parameter in the determination of annoyance. Moreover, sound emergence does not provide reliable information about the audibility of the specific sound. From a practical perspective the measurement of emergence raises several issues relating to the understanding of the specific sound, the access to the background sound and measurement uncertainty. In addition, sound emergence is problematic from a planning perspective because this indicator relies on the shifting baseline of residual sound. Therefore sound emergence offers little guarantee of compliance to the different stakeholders. Furthermore, the necessity to predict background sound makes the simulation of emergence in software more challenging and uncertain than the one of the sound pressure level from a specific source.

From all this it would seem reasonable to reconsider the use of sound emergence in legislations that rely on it. The weight of history is not enough and research is needed to provide evidence that sound emergence is relevant for setting environmental noise limits. Further research about the potential correlation between annoyance and sound emergence from specific sources is necessary. Due to the rapid and global development of wind energy, the justification of the use of sound emergence in the environmental impact assessment of wind farms is certainly worth investigating. Since sound emergence appears to be a poor predictor of audibility, the temporal and spectral characteristics of the specific source should be taken into account in the estimation of sound emergence leading to source-specific choices of the metrics used for estimating total and residual sound pressure levels.

If switching to another metric is out of question we can make the following recommendations: (1) the specific sound should be defined from the source perspective, (2) the metrics used for residual and specific sound should be specified unambiguously, (3) the estimation of emergence should be based on long term measurements to account for the variability of the residual and the specific sound and (4) sound emergence should be used in combination with limits based on sound levels to avoid the shifting baseline phenomenon. Attempts have been already made in this direction but they are only incompletely successful.

In our opinion and in the wake of [35], combining (i) an estimation of the audibility of the specific sound for the community and (ii) an estimation of L_{spec} would be by far superior to sound emergence while serving the same purposes. The audibility assessment could be built on the above-mentioned previous research [41]. Regarding L_{spec}, this quantity should be obtained with a sufficient signal-to-noise ratio. This may require the acceptance that source-attributable noise at the receiver may not always be accessible to direct measurements of sound pressure level and that simulations may be necessary.

Author Contributions: conceptualization, G.D.; methodology, G.D.; investigation, G.D. and T.G.; resources, G.D., G.L. and T.G.; data curation, G.D. and T.G.; writing—original draft preparation, G.D.; writing—review and editing, G.D., G.L. and T.G.; G.D. carried out the literature search and analysis, analyzed the connection between sound emergence and signal-to-noise ratio, and wrote the first draft, T.G. contributed to the literature search, contributed to the discussion of the connection between annoyance and sound emergence and reviewed the first draft. G.L. contributed to the literature search, reviewed the first draft and contributed to it.

Funding: This research received no external funding.

Acknowledgments: The first author wishes to thank the members of French AFNOR/S30J environmental noise standardization committee for the numerous discussions that helped him identify the different issues brought by sound emergence when used outdoors.

Conflicts of Interest: The authors declare no conflict of interest.

References

1. Schafer, R.M. *The Soundscape—Our Sonic Environment and the Tuning of the World*, 2nd ed.; Destiny Books: Rochester, VT, USA, 1977.
2. I-INCE. *Survey of Legislation, Regulations, and Guidelines for Control of Community Noise*; Report 09-1; International Institute of Noise Control Engineering: Petaluma, CA, USA, 2009.
3. ISO. *Acoustics, Assessment of Noise with Respect to Community Response*; Standard ISO/R1996; International Organization for Standardization: Geneva, Switzerland, 1971.
4. ISO. *Acoustics, Descrption, Measurement and Assessment of Environemental Noise—Part 1: Basic Quantities and Assessment Procedures*; Standard ISO 1996-1:2016; International Organization for Standardization: Geneva, Switzerland, 2016.
5. Fields, J.M. Effect of personal and situational variables on noise annoyance in residential areas. *J. Acoust. Soc. Am.* **1993**, *93*, 2753–2763. [CrossRef]
6. AFNOR. *Acoustique, Caractérisation et Mesurage des Bruits dans L'environnement, Méthode Particulière de Mesurage*; Standard NF S 31-010; Association Française de Normalisation: Saint-Denis, France 1996.
7. ISO. *Quantities and Units, Part 8:Acoustics*; Standard ISO 80000-8:2007; International Organization for Standardization: Geneva, Switzerland, 2007.
8. World Bank. *Environmental, Health, and Safety General Guidelines (English)*; Number 112110, International Finance Corporation E&S; World Bank Group: Washington, DC, USA, 2007.
9. World Bank. *Environmental, Health, and Safet Guidelines for Wind Energy*; Number 110346, International Finance Corporation E&S; World Bank Group: Washington, DC, USA, 2007.
10. Premier Ministre. Arrêté du 21 novembre 2017 relatif aux prescriptions générales applicables aux installations classées pour la protection de l'environnement soumises à déclaration sous la rubrique n° 2793-3a. *J. Off. De La République Française* **2017**, *0281*, 47–53.
11. Comission technique d'étude du bruit. Avis du 21 juin 1963 concernant l'estimation des troubles produits par l'excès de bruit. *J. Off. De La République Française* **1963**.
12. Premier Ministre. Arrêté du 26 août 2011 relatif aux installations de production d'électricité utilisant l'énergie mécanique du vent au sein d'une installation soumise à autorisation au titre de la rubrique 2980 de la législation des installations classées pour la protection de l'environnement. *J. Off. De La République Française* **2011**, *0198*, 57–62.
13. AFNOR. *Acoustique—Mesurage du Bruit dans L'environnement avant et Après Installation Éolienne*; Draft Standard PR NF S31-114; Association Française de Normalisation: Saint-Denis, France, 2011.
14. Premier ministre. Décret n° 2006-1099 du 31 août 2006 relatif à la lutte contre les bruits de voisinage et modifiant le code de la santé publique (dispositions réglementaires). *J. Off. De La République Française* **2006**, *202*, 13042.
15. ISO. *Acoustics, Objective Method for Assessing the Audibility of Tones in Noise, Egineering Method*; Standard ISO/PAS 20065:2016; International Organization for Standardization: Geneva, Switzerland, 2016.
16. Premier Ministre. Décret n° 2017-1244 du 7 août 2017 Relatif à la Prévention des Risques Liés aux Bruits et aux Sons Amplifiés. *J. Off. De La République Française* **2017**, *0185*, 101–105.
17. Presidente del Consiglio dei Ministri. Decreto, Limiti massimi di esposizione al rumore negli ambienti abitativi e nell'ambiente esterno. *Gazz. Uff.* **1991**, *57*.
18. Presidente del Consiglio dei Ministri. Decreto, Determinazione dei valori limite delle sorgenti sonore. *Gazz. Uff.* **1997**, *280*.
19. Ministero dell'Ambiente e della Tutela del Territorio. Circolare 6 settembre 2004—Interpretazione in materia di inquinamento acustico: Criterio differenziale e applicabilità dei valori limite differenziali. *Gazz. Uff.* **2004**, *217*.
20. Gallo, P.; Fredianelli, L.; Palazzuoli, D.; Licitra, G.; Fidecaro, F. A procedure for the assessment of wind turbine noise. *Appl. Acoust.* **2016**, *114*, 213–217. [CrossRef]
21. Ministero dell'Ambiente. Decreto, Tecniche di rilevamento e di misurazione dell'inquinamento acustico. *Gazz. Uff.* **1998**, *76*.

22. Schild, J.; Chavand, V. State of the art and the new perspective for the development of noise regulation of wind farms. In *Proc Wind Turbine Noise*; INCE Europe, Southport, UK, 2015.

23. Koppen, E.; Fowler, K. International legislation for wind turbine noise. In *Proceedings of the Euronoise*; NAG, ABAV: Maastricht, The Netherlands, 2015; pp. 321–326.

24. Kelly, D.; Dilworth, C. *Guidance Note for Noise: Licence Applications, Surveys and Assessment in Relation to Scheduled Activities (NG4)*; Technical Report; Environmental Protection Agency—Office of Environment Enforcement: Wexford, Ireland, 2016.

25. McAleer, S.; McKenzie, A. *Guidance Note on Noise Assessment of Wind Turbine Operations at EPA Licensed Sites (NG3)*; Technical Report; Environmental Protection Agency—Office of Environment Enforcement: Wexford, Ireland, 2011.

26. BSI. *Methods for Rating and Assessing Industrial and Commercial Sound*; Standard BS 4142:2014; British Standards Institution: London, UK, 2014.

27. ETSU. *The Assessment and Rating of Noise from Wind Farms*; Technical Report R97; Energy Technology Support Unit: Harwell, UK, 1997.

28. Burgess, M.; Renew, W. Overview of environmental noise policies in Australia. *Acoust. Aust.* **2003**, *31*, 5–9.

29. EPA New South Wales. *Noise Guide for Local Goverment*; Environment Protection Authority: Sydney, Australia, 2013.

30. EPA Southern Australia. *Wind Farm Environmental Noise Guideline*; Environment Protection Authority: Adelaide, Australia, 2009.

31. EPA Victoria. *Noise Control Guidelines*; Technical Report 1254; Environemntal Pollution Authority: Melbourne, Australia, 2008.

32. Environment Division. *Noise Measurement Procedures*, 2nd ed.; Department of Environment, Parks, Heritages and the Arts: Hobart, Australia, 2008.

33. EPA New South Wales. *Draft Industrial Noise Guideline*; Environment Protection Authority: Sydney, Australia, 2015.

34. Stallen, P.J. A theoretical framework for environmental noise annoyance. *Noise Health* **1999**, *1*, 69–80. [PubMed]

35. Aubrée, D.; Rapin, J.M. *Les Indices de Bruit et la Gêne*; Technical Report GA 712/92-192/DAJMR/YC; CSTB: Grenoble, FR, 1993.

36. Viollon, S.; Marquis-Favre, C.; Junker, F.; Baumann, C. Environmental assessment of industrial noises annoyance with the criterion "sound emergence". In Proceedings of the ICA, Kyoto, Japan, 4–9 April 2004.

37. Rumeau, M. Etude d'un indicateur de gêne due aux bruits impulsionnels. *Acta Acust.* **1994**, *2*, 497–504.

38. Berry, B. Recent advances in the measurement and rating of impulsive noise. In *Proceedings of the 13th International Congresses on Acoustics*; Sava Centar: Belgrade, Republika Srbija, 1989; Volume 3, pp. 147–150.

39. Fields, J.M. Impact of ambient noise on noise annoyance: An assessment of the evidence. In *INTER-NOISE and NOISE-CON Congress and Conference Proceedings*; Daigle, G.A., Stinson, M.A., Eds.; Noise Control Fundation: Toronto, ON, Canada, 1992.

40. Bütikofer, R.; Eggenschwiler, K.; Steiner, E. Annoying low level sound. In Proceedings of the ICBEN, Zürich, Switzerland, 18–22 June 2017; p. 6.

41. Fidell, S.; Teffeteller, S.; Horonjeff, R.; Green, D. Predicting annoyance from detectability of low-level sounds. *J. Acoust. Soc. Am.* **1979**, *66*, 1427. [CrossRef]

42. Fidell, S.; Pearsons, K.S.; Bennet, R. Prediction of aural detectability of noise signals. *Hum. Factors* **1974**, *16*, 373–383. [CrossRef] [PubMed]

43. Bolin, K.; Kedhammar, A.; Nilsson, M.E. The influence of background sound on loudness and annoyance of wind turbine noise. *Acta Acust. United Acust.* **2012**, *98*, 741–748. [CrossRef]

44. Van Renterghem, T.; Bockstael, A.; De Weirt, V.; Botteldooren, D. Annoyance, detection and recognition of wind turbine noise. *Sci. Total. Environ.* **2013**, *456–457*, 333–345. [CrossRef]

45. Miedema, H.M.E.; Vos, H. Noise annoyance from stationary sources: Relationships with exposure metric day–evening–night level (DENL) and their confidence intervals. *J. Acoust. Soc. Am.* **2004**, *116*, 334–343. [CrossRef]

46. Joint Committee for Guides in Metrology WG1. *Evaluation of Measurement Data—Guide to the Expression of Uncertainty in Measurement (GUM)*; Standard JCGM 100; BIPM: Sèvres, France, 2008.

47. IEC. *Electroacoustics—Sound Level Meters—Part 1: Specifications*; Standard IEC 61672-1; International Electrotechnical Commission International Electrotechnical Commission: Geneva, Switzerland, 2013.

48. Casini, D. From standard specifications to L_{Aeq} uncertainty in environmental noise measurements. *Acta Acust.* **2018**, *104*, 707–719. [CrossRef]

49. Vincent, E.; Virtanen, T.; Gannot, S. (Eds.) *Audio Source Separation and Speech Enhancement*; Wiley: Hoboken, NJ, USA, 2018.

50. Fredianelli, L.; Gallo, P.; Licitra, G.; Carpita, S. Analytical assessment of wind turbine noise impact at receiver by means of residual noise determination without the windfarm shutdown. *Noise Control Eng. J.* **2017**, *65*, 417–433. [CrossRef]

51. ISO. *Acoustics, Normal Equal-Loudness-Level Contours*; Standard ISO 226:2003; International Organization for Standardization, Geneva, Switzerland, 2003.

52. Fastl, H.; Zwicker, E. *Psychoacoustics: Facts and Models*; Springer: Berlin/Heidelberg, Germany, 2007; Volume 22, pp. 1–463.

53. Daniel, E. Noise and Hearing Loss: A Review. *J. Sch. Health* **2007**, *77*, 225–231. [CrossRef] [PubMed]

54. Pauly, D. Anecdotes and the shifting baseline syndrome of fisheries. *Trends Ecol. Evol.* **1995**, *10*, 430. [CrossRef]

55. De Coensel, B.; Bottledooren, D.; De Muer, T. 1/f noise in rural and urban soundscapes. *Acta Acust. United Acust.* **2003**, *89*, 287–295.

 International Journal of
Environmental Research and Public Health

Communication

A Taxonomy Proposal for the Assessment of the Changes in Soundscape Resulting from the COVID-19 Lockdown

César Asensio [1],*, Pierre Aumond [2], Arnaud Can [2], Luis Gascó [1], Peter Lercher [3], Jean-Marc Wunderli [4], Catherine Lavandier [5], Guillermo de Arcas [1], Carlos Ribeiro [6], Patricio Muñoz [7] and Gaetano Licitra [8],*

[1] Instrumentation and Applied Acoustics Research group (I2A2), Universidad Politécnica de Madrid, 28031 Madrid, Spain; luisgascosanchez@gmail.com (L.G.); g.dearcas@upm.es (G.d.A.)
[2] UMRAE, Univ Gustave Eiffel, IFSTTAR, CEREMA, 44340 Bouguenais, France; pierre.aumond@univ-eiffel.fr (P.A.); arnaud.can@ifsttar.fr (A.C.)
[3] Institute for Highway Engineering and Transport Planning, Graz University of Technology, 8010 Graz, Austria; peter.lercher.at@gmail.com
[4] Empa, Swiss Federal Laboratories for Material Science and Technology, Laboratory for Acoustics/Noise Control, 8600 Dübendorf, Switzerland; jean-marc.wunderli@empa.ch
[5] ETIS Laboratory, UMR 8051, CY Cergy Paris University, ENSEA, CNRS, F-95302 Cergy-Pontoise Cedex, France; catherine.lavandier@cyu.fr
[6] Bruitparif, 93200 Saint-Denis, France; Carlos.Ribeiro@bruitparif.fr
[7] Acoucite, Observatoire de l'environnement sonore de la Métropole de Lyon, 69007 Lyon, France; patricio.munoz@acoucite.org
[8] Environmental Protection Agency of Tuscany Region, Pisa Department, 56127 Pisa, Italy
* Correspondence: casensio@i2a2.upm.es (C.A.); g.licitra@arpat.toscana.it (G.L.)

Received: 16 May 2020; Accepted: 5 June 2020; Published: 12 June 2020

Abstract: Many countries around the world have chosen lockdown and restrictions on people's mobility as the main strategies to combat the COVID-19 pandemic. These actions have significantly affected environmental noise and modified urban soundscapes, opening up an unprecedented opportunity for research in the field. In order to enable these investigations to be carried out in a more harmonized and consistent manner, this paper makes a proposal for a set of indicators that will enable to address the challenge from a number of different approaches. It proposes a minimum set of basic energetic indicators, and the taxonomy that will allow their communication and reporting. In addition, an extended set of descriptors is outlined which better enables the application of more novel approaches to the evaluation of the effect of this new soundscape on people's subjective perception.

Keywords: COVID-19; noise; soundscape; metrics; indicators; descriptors; sound; lockdown

---

1. Introduction

Unfortunately, the year 2020 will be known as the year of the coronavirus disease (COVID-19) pandemic. To a greater or lesser extent, the epidemic has spread to every continent, without distinction, affects all ages and is particularly dangerous for older people. The strategies designed by different governments to combat the pandemic in many countries have been very diverse, but many countries have chosen lockdown and restrictions on people's mobility [1]. More than 3.9 billion people, or a half of the world's population living in 90 different countries around the world have been under containment as a measure to maintain social distancing [2].

Commercial flights, both international and domestic, have been severely restricted, with all flights not dedicated to the provision of medical supplies and other essential products being affected in many countries [3]. Likewise, ground transportation has also been severely restricted, with substantial percentages of the population unable to access their jobs or having to work remotely [4].

In addition to the dreadful consequences that the pandemic has had on the population, in terms of infections, hospitalizations and the number of deaths, the lockdown of people and their absence from the environment has had considerable environmental consequences, with animal species returning to the urban environment and beaches, and varying reductions in peak and average air pollution levels in populated areas [5–7].

As a result of restrictions on urban mobility, traffic noise has been drastically reduced. Conversely, natural noise, such as bird singing, is emerging again, although it is difficult to know whether this is related to a closer presence of the source, an increase in levels, lack of masking noise or a perceptual effect, and whether it is due to the lockdown or not [8].

Therefore, the acoustics community has been mobilized. National acoustical associations in Italy and UK launched initiatives to collect measurement campaign data [9,10] and many consultants, engineers, research groups and noise management authorities around the world have begun to produce reports to address, through measurement data, the assessment of the reduction that confinement has produced in the environmental noise of each city. Although a few of the initiatives gave some general indications, there is a risk that these interesting reports, coming from personal and structured actions, suffer from a lack of consistency that makes it almost impossible to compare them, which would be extremely challenging for the overall analysis of the effect on the confinement on human behaviors and perception.

At the same time, new projects are active to collect recordings and metadata of sounds in the COVID-19 scenario, such as the LYS (locate your sound) project [11] in Italy with around 3000 recordings on 6 May 2020, showing the richness of lived experiences and the value of the recordings so that people do not forget and recover lost sounds. Also, through sound recording and automatic audio tagging of recordings, the Silent-Cities Project aims to create a database of audio files that allows to study, among other things, the relationship between natural and human-generated sounds in different levels of economic activity [12]. Also related to this topic, Acoucité has developed a questionnaire oriented towards assessing population feelings about the changes in the noise environment since lockdown [13].

Since it is expected that in the coming months these preliminary analyses will become scientific articles, it is considered very necessary to establish a common framework to harmonize the basic results of these investigations, so that comparisons can be made between different populations and countries, leading to a macro-analysis that will make it possible to know and evaluate the overall effect of confinement, to compare the effect of different confinement strategies and to communicate this information to the public.

In order to achieve these objectives, in this communication we propose a minimum set of common descriptors, which will make it possible to assess noise pollution in each location, and to appraise the noise reduction that the measures against COVID-19 imply. In addition, to give a status of open data to all this information, and to facilitate future analyses, we propose a data structure that gathers all the noise-related data information in the form of a taxonomy. Although this data structure arises as a necessity for the comparison of noise studies related to COVID-19 effects, it should also be valid for the assessment of noise in the future, with minor changes both in exceptional and everyday circumstances.

2. Noise Descriptors and Taxonomy for Physical Characterization

This paper focuses on indicators for physical characterization of noise, since an important part of the analysis will probably deal with the pre-post comparison based on the noise monitoring systems implemented in cities and airports. Indicators that aim to assess people's exposure to noise are widespread [14]. With their benefits and shortcomings, they allow a description to be made based on objective criteria, such as the acoustic energy contained in the environment.

2.1. Measurement Data Structure

We recommend that each measurement be described by the following set of data, which will refer to a time interval starting at the day and time referenced. We propose to use a simple open file format such as the comma-separated values (CSV) file to share the raw data. The field names of the first row of the dataset are shown in Table 1, and each row of the file will describe a measurement.

It is recommended that in each location, the basic data set reported on a daily basis be L_n and L_{den}, following the recommendations of the Environmental Noise Directive [15]. Additionally, it is considered convenient to add, if available, as an extended data set, the time series of measurements of equivalent sound level of one hour (either A or Z weighted, $L_{Aeq,1h}$, $L_{eq,1h}$). Therefore, this recommendation includes 24 descriptors a day (24 $L_{Aq,1\,h}$ or $L_{eq,1\,h}$ values). The same data structure can be valid for daily or hourly basis, using the time of indicator definition, duration and starting time.

Table 1. Measurement data structure.

Field	Description	Data Type
Identification	Short name, to identify the measurement location	String
City	City	String
Country	Country	String
Measurement provider	Entity that is providing the measurements (i.e., local authority or airport manager)	String
Coordinates	Measurement location, WGS84 format	String latitude, longitude "48.856614; 2.3522219"
Instrument class	Certified instruments should be considered, either type 1 or 2. Non-certified (but calibrated) sensors, type 3	Integer (1, 2, 3)
Instrument brand	Type of area (residential, hospital, school, ...)	String
Prevailing sound sources	Semicolon delimited tags to describe the area, showing the prevailing sound sources	String (road, air, rail, nightlife, etc.)
Date/Time	Measurement starting date and o'clock time	String YYYYmmddThh0000
Stage	Before lockdown = 1 Lockdown = 2 After lockdown = 3	Integer (1, 2, 3)
Description of the stage	A qualitative description of the period to analyze. It will be used to understand the level of lockdown in the city where the measurements were taken. Some tags are proposed.	String. Using tags: (a) events suspended; (b) schools closed; (c) non-essential shops closed; (d) non-essential movement banned; (e) land border closed; (f) non-essential production closed [16]
Duration	Measurement duration. Only necessary for indicator type L_{eq}.	Integer (minutes)
Indicator	Type of indicator	String (L_{eq}, L_{den}, L_n...)
Frequency weighting	Frequency weighting	String (A, Z)
Measurement	The value of the indicator	Float, 1 decimal digit (decibel)
Miscellaneous	Free comment about the data collection	String

It is necessary to ensure the reliability of the data, so that measurements that could be affected by weather, maintenance operations or unusual sound events, that could affect the measurements, are excluded.

Table S1, provided as supplementary material, contains an example of a data file, according to this measurement data structure.

2.2. Data to Report

For data processing and reporting, local diversities and uses may result in large differences that prevent comparison of results. Each study can have a very different scope and objectives, and thus the results reported can vary considerably. However, we consider that analyzing the reduction of noise produced during lockdown may be an objective common to all of them, and, focusing on the evaluation of such reduction, we propose a series of indicators that may be useful, considering them as a set of minimums that all studies should address. For this reason, we recommend that the reports contain, at least, a time series (chart or table) for L_{den} and L_n, and the information specified in Table 2:

Table 2. Minimums to report.

Measurement Location:	Identification	STAGE		
		Before	Lockdown	After
Working day	% days exceeding $L_{den} = 65$			
	% days exceeding $L_n = 55$			
	Average $L_{Aeq,1\,h}$ during rush hour (dBA)			
	Average $L_{Aeq,1\,h}$ during off-peak hour (dBA)			
	Average L_{den} (dBA)			
	Average L_n (dBA)			

Notes: Arithmetic averages must be considered. The "Before" stage is the one that determines rush and off-peak hour. It will be different for working days and weekends.

In some locations, due to their characteristics, it may be of interest to evaluate the reduction occurring during weekends or holidays. In this case, the information contained in Table 2 can be replicated, redefining the peak and valley hours, depending on the prevailing noise source in each area.

2.3. Data Collection

Although data collection is out of the scope of this communication, we encourage providers to share their database with the community on the Zenodo platform, which is an open-access repository operated by CERN [17]. For each submission, a persistent digital object identifier (DOI) is given, which makes the stored items easily citable. The upload limit is about 50 GB. To identify all the databases that will have followed the protocol recommended in this communication. Please add the tags: "Noise", "COVID-19"; "Lockdown", "Taxonomy".

3. Extended Indicators

The previous section focuses on describing the noise dose, and how it has decreased because of the reduced mobility and human activities that confinement has produced. This is an aspect that has been well studied over decades, so it has been relatively easy to agree on a set of data, which we believe noise monitoring systems will be recording on a regular basis.

However, this set of indicators does not fully describe the subjective experience that the new soundscape draws. Sudden shift in sound environments include changes in noise dynamics, and the emergence of unusual sound sources. Beyond the purely energetic effect that derives from the confinement, it is foreseeable that the perception of change in the soundscape will be different according to cultural aspects [18,19]. This can only be widely investigated if an adequate set of descriptors, conveniently harmonized at international level, are defined. This requires an extended set of indicators needed for more detailed analyses.

These types of investigations are not so widespread in the different areas of noise management in public administrations, and therefore there are restrictions with respect to the technical knowledge of the staff who must carry out the measurements. This is the reason why we wanted to include a classification of indicators in this paper, that may be helpful for future research, and which may still be used to describe outdoor sound in the face of the unique phenomenon we are experiencing, from different points of view, such as biophony or soundscape.

Table 3 also includes the energetic indicators already mentioned in the previous section, to give consistency, and to allow comparison of the different types of noise descriptors. The following indicators should be calculated on an hourly basis.

Table 3. Extended indicators.

	Indicators and Description	Physical Descriptive Power	Perceptive Descriptive Power
Energetic indicators	L_{eqT} continuous equivalent sound pressure level during time period T L_n continuous equivalent sound pressure level during night period L_{den}, day, evening, night combined indicator [20–22]	Cumulative energetic indicators. A, C or Z frequency weighting	Correlated to long term health effects
Statistical indicators	L_{90} [23], 90% percentile level	Describes background noise	Does not emerge from studies
	L_{50}, 50% percentile level [24]	Good for discriminating sound environments	Very good correlation with perceived sound intensity and sound pleasantness
	L_{10}, 10% percentile level [23–25]	Describes contribution of loudest events	Outperforms L_{Aeq} to describe perception of high noise levels
Spectrum and source related indicators	Sound ecology indicators: NDSI, normalized difference soundscape index; ACI, acoustic complexity in; entropy; BIO, bioacoustic index; ADI, acoustic diversity index; AEI, acoustic evenness index [11,26]	Good for discriminating presence of biophonic sounds and anthropogenic sounds in urban sound	Likely to be correlated with the time presence of the described sound sources
	The normalized time and frequency second derivative: $TFSD_{mean, 4k\ Hz}$ (birds); $TFSD_{mean, 500\ Hz}$ (human voices) [27,28]	Can be computed from octave band 1 s dataset. Good for discriminating presence of biophonic sounds and anthropogenic sounds in urban sound environment	Likely to be correlated with the time presence of the described sound sources
	L_{eq} (63 Hz–500 Hz); 1/3 octave band continuous sound pressure level [28,29]	Good for discriminating sound environments frequency content	Correlated with the time presence of Traffic
	L_{Ceq}-L_{Aeq}, difference between A- and C-weighted equivalent continuous sound levels [30–34]	Describes the amount of low frequencies	Differences of 15 to 20 dB show an effect on annoyance and perception of vibrations
Emergences and noise variation indicators	L_{Amax}, maximum A-weighted noise level; NA, number of events above a threshold; time above a threshold [35,36]	NA80, number of events above a 80 dBA, or TA80 time above 80 dBA (additional thresholds can be considered)	Awakening probability with increasing L_{Amax} The number of high noise level events may affect sleep motility. For aircraft noise, also an effect on annoyance is suggested
	Calculated from percentiles. Fluctuation: defined as the difference between the (single) source event and the source background level. Emergence: Difference between the source event and the overall background level (L10–L90 or L1–L99) [37–42]	Good description of the energetic increase produced by a source	Field investigations on annoyance and hypertension yield some support in the context of mixed sound exposure and low background levels (main roads). No consensus concerning the perceptive effects
	Intermittency ratio (IR). Ratio between the sound energy contributions of events, and the overall contributions during the measurement period [43–46]	Expresses the energetic share of noise exposure created by individual noise events	Highly intermittent nocturnal noise is correlated with increased risk of cardiovascular diseases. In a fully adjusted hypertension model the IR made an additional contribution beyond the L_{den} in mixed source exposure situations. IR has an additional effect on %HA and can explain shifts of the exposure-response curve of up to about 6 dB.

4. Conclusions

The COVID-19 pandemic has significantly modified urban sound environments, opening up an unprecedented opportunity for research in the field. In order to enable these investigations to be carried out in a more harmonized and consistent manner, the group of experts implied in this article agreed on a minimum set of indicators that should imperatively be calculated. Recommendations are also given as concerning the measurement data structure (taxonomy) for the global assessment of the effect that the lockdown due to COVID-19 has produced on environmental noise.

Beyond this minimum, the selection of a set of descriptors that are capable of adequately describing citizens' perception of any new circumstance would be highly desirable, to serve as a guide for future research. For this reason, an overview of an extended set of indicators is presented. These indicators cover all the physical dimensions of sound environments, and are supported by elements of literature: Energetic, spectral and temporal dimension, emergence and source-related indicators. Thus, this extended set of indicators should allow a more detailed analysis of the changes in noise environments related to confinement, and to a broader extent help in understanding the impact on sound environments of any policy achieved at the urban scale.

Finally, the COVID-19 crisis has revealed a big lack in the current state-of-the-art to analyze urban sound environments. The noise indicators mainly deal with sound environments as a whole, and do not distinguish between the sound sources that compose it. The sound environments introduced by the lockdowns modified them not only in levels, but also by the present sources. Natural sounds are heard again, both because there is less noise to mask them, and because of the reappearance of animal species in areas usually occupied by vehicles and people. In these circumstances, even the sounds that were previously integrated to form our acoustic environment now, in isolation, acquire a very particular character, and may be especially relevant. When the passage of a vehicle was hidden by the noise of traffic as a whole, now the movement of each vehicle acquires a whole different meaning. Not to mention other sounds, such as the passing of ambulances, which in the pandemic may intensify their meaning and fully change people's perception.

This dimension is unfortunately absent from current indicators. Therefore, the development of source-orientated indicators, able to quantify the presence of sources of interest, and ideally performing with urban sound mixtures with strong temporal overlaps, is strongly advocated. Premises towards such indicators can be found in the literature, relying on sound recognition [25,47–49].

The physical indicators proposed, although they are linked to perceptual and health effects, will most likely be insufficient to capture the entire sound experience. Sensitive data, such as the speed of the experienced change, the link that can exist between the sound environment and its emotional evocation, the diversity in the life situations of city dwellers faced with the lockdown, cannot be captured by physical indicators. They are, however, still an integral part of the soundscapes during this period. Although emphasized in this specific period, this lack stands for any observed modification in sound environments. This advocates for the collection of sensitive data, in addition to physical data, as part of the next generation of measurement networks [49,50].

Supplementary Materials: The following are available online at http://www.mdpi.com/1660-4601/17/12/4205/s1, Table S1: Data file example.

Author Contributions: Conceptualization, C.A., G.L.; writing original draft preparation, C.A.; state of art: all authors; writing, review, and editing: C.A., G.L., L.G., P.A. A.C., J.-M.W.; review: C.L., P.L., G.d.A., C.R., P.M. All authors have read and agreed to the published version of the manuscript.

Funding: This research received no external funding.

Conflicts of Interest: The authors declare no conflicts of interest.

References

1. Aktay, A.; Bavadekar, S.; Cossoul, G.; Davis, J.; Desfontaines, D.; Fabrikant, A.; Gabrilovich, E.; Gadepalli, K.; Gipson, B.; Guevara, M.; et al. Google COVID-19 Community Mobility Reports: Anonymization Process Description (version 1.0). *arXiv* **2020**, arXiv:2004.04145.
2. Coronavirus: Half of Humanity Now on Lockdown as 90 Countries Call for Confinement. Available online: https://www.euronews.com/2020/04/02/coronavirus-in-europe-spain-s-death-toll-hits-10-000-after-record-950-new-deaths-in-24-hou (accessed on 13 May 2020).
3. Olive, X.; Strohmeier, M.; Lübbe, J. Crowdsourced air traffic data from The OpenSky Network 2020. *OpenAIRE* **2020**. [CrossRef]
4. Tardivo, A.; Martín, C.S.; Zanuy, A.C. Covid-19 impact in Transport, an essay from the Railways' system research perspective. *Pract. Pipeline* **2020**. Available online: https://advance.sagepub.com/articles/Covid-19_impact_in_Transport_an_essay_from_the_Railways_system_research_perspective/12204836 (accessed on 15 May 2020). [CrossRef]
5. IQAir. *IQAir COVID-19 Air Quality Report*; IQAir: Goldach, Switzerland, 2020.
6. Venter, Z.S.; Aunan, K.; Chowdhury, S.; Lelieveld, J. COVID-19 lockdowns cause global air pollution declines with implications for public health risk. *medRxiv* **2020**. Available online: https://www.medrxiv.org/content/early/2020/04/14/2020.04.10.20060673 (accessed on 15 May 2020). [CrossRef]
7. Monitoring Covid-19 Impacts on Air Pollution. Available online: https://www.eea.europa.eu/themes/air/air-quality-and-covid19/monitoring-covid-19-impacts-on (accessed on 13 May 2020).
8. Sounds from the Global Covid-19 Lockdown. Available online: https://citiesandmemory.com/covid19-sounds/ (accessed on 13 May 2020).
9. Grande Partecipazione all'iniziativa AIA di caratterizzazione dei Livelli Sonori Durante l'emergenza da Coronavirus. Available online: https://acustica-aia.it/grande-partecipazione-alliniziativa-aia-di-caratterizzazione-dei-livelli-sonori-durante-lemergenza-da-coronavirus/ (accessed on 13 May 2020).
10. COVID-19: The Quiet Project—Call for Measurements. Available online: https://www.ioa.org.uk/news/covid-19-quiet-project-%E2%80%93-call-measurements (accessed on 13 May 2020).
11. Locate Your Sound—Paesaggi Sonori Italiani #Covid19. Available online: https://locateyoursound.com/en/ (accessed on 13 May 2020).
12. Challéat, S.; Farrugia, N.; Gasc, A.; Froidevaux, J.; Hatlauf, J.; Dziock, F.; Charbonneau, A.; Linossier, J.; Watson, C.; Ullrich, P.A. Silent·Cities. **2020**. Available online: osf.io/h285u (accessed on 15 May 2020). [CrossRef]
13. Confinement COVID-19: Impact sur l'environnement Sonore. Available online: http://www.acoucite.org/confinement-covid-19-impact-sur-lenvironnement-sonore/ (accessed on 13 May 2020).
14. Can, A.; Aumond, P.; Michel, S.; de Coensel, B.; Ribeiro, C.; Botteldooren, D.; Lavandier, C. Comparison of noise indicators in an urban context, Inter-Noise 2016. In Proceedings of the 45th International Congress and Exposition of Noise Control Engineering, Aughambourg, Germany, 21–24 August 2016; p. 9.
15. *European Parliament Directive 2002/49/EC of the European Parliament and of the Council of 25 June 2002 Relating to the Assessment and Management of Environmental Noise*; European Parliament and Council: Brussels, Belgium, 2002.
16. Europe's Coronavirus Lockdown Measures Compared. Available online: https://www.politico.eu/article/europes-coronavirus-lockdown-measures-compared/ (accessed on 15 May 2020).
17. Sicilia, M.; García-Barriocanal, E.; Sánchez-Alonso, S. Community Curation in Open Dataset Repositories: Insights from Zenodo. *Procedia Comput. Sci.* **2017**, *106*, 54–60. [CrossRef]
18. Yu, C.; Kang, J. Soundscape in the sustainable living environment: A cross-cultural comparison between the UK and Taiwan. *Sci. Total Environ.* **2014**, *482*, 501–509. Available online: http://www.sciencedirect.com/science/article/pii/S0048969713012515 (accessed on 15 May 2020). [CrossRef]
19. Sato, T.; Yano, T.; Björkman, M.; Rylander, R. Comparison of community response to road traffic noise in japan and sweden—Part i: Outline of surveys and dose–response relationships. *J. Sound Vib.* **2002**, *250*, 161–167. Available online: http://www.sciencedirect.com/science/article/pii/S0022460X01938921 (accessed on 15 May 2020). [CrossRef]
20. European Comission. *European Comission Position Paper on EU Noise Indicators*; European Comission: Brussels, Belgium, 2000.

21. Gozalo, G.R.; Carmona, J.T.; Morillas, J.B.; Vílchez-Gómez, R.; Escobar, V.G. Relationship between objective acoustic indices and subjective assessments for the quality of soundscapes. *Appl. Acoust.* **2015**, *97*, 1–10. [CrossRef]

22. Alberola, J.M.; Flindell, I.H.; Bullmore, A.J. Variability in road traffic noise levels. *Appl. Acoust.* **2005**, *66*, 1180–1195. [CrossRef]

23. Nelson, P. *Transportation Noise Reference Book*; Butterworths: London, UK, 1987.

24. Can, A.; Gauvreau, B. Describing and classifying urban sound environments with a relevant set of physical indicators. *J. Acoust. Soc. Am.* **2015**, *137*, 208–218. [CrossRef] [PubMed]

25. Axelsson, O.; Nilsson, M.E.; Berglund, B. A principal components model of soundscape perception. *J. Acoust. Soc. Am.* **2010**, *128*, 2836–2846. [CrossRef] [PubMed]

26. Gontier, F.; Lavandier, C.; Aumond, P.; Lagrange, M.; Petiot, J.F. Estimation of the perceived time of presence of sources in urban acoustic environments using deep learning techniques. *Acta Acust. United Acust.* **2019**, *105*, 1053–1066. [CrossRef]

27. Aumond, P.; Can, A.; Coensel, B.D.; Botteldooren, D.; Ribeiro, C.F.; Lavandier, C. Modeling soundscape pleasantness using perceptual assessments and acoustic measurements along paths in urban context. *Acta Acust. United Acust.* **2017**, *103*, 430–443. [CrossRef]

28. Torija, A.J.; Ruiz, D.P.; Ramos-Ridao, A.F. Application of a methodology for categorizing and differentiating urban soundscapes using acoustical descriptors and semantic-differential attributes. *J. Acoust. Soc. Am.* **2013**, *134*, 791–802. [CrossRef]

29. Ascari, E.; Licitra, G.; Teti, L.; Cerchiai, M. Low frequency noise impact from road traffic according to different noise prediction methods. *Sci. Total Environ.* **2015**, *505*, 658–669. [CrossRef]

30. Kjellberg, A.; Tesarz, M.; Holmberg, K.; Landström, U. Evaluation of frequency-weighted sound level measurements for prediction of low-frequency noise annoyance. *Environ. Int.* **1997**, *23*, 519–527. [CrossRef]

31. Cik, M.; Lienhart, M.; Lercher, P. Analysis of Psychoacoustic and Vibration-Related Parameters to Track the Reasons for Health Complaints after the Introduction of New Tramways. *Appl. Sci.* **2016**, *6*, 398. [CrossRef]

32. Leventhall, H. Low frequency noise and annoyance. *Noise Health* **2004**, *6*, 72.

33. DIN. *DIN 45680: 2013 Measurement and Assessment of Low-frequency Noise Immissions in the Neigbourhood*; DIN: Berlin, Germany, 2013.

34. Janssen, S.A.; Centen, M.R.; Vos, H.; van Kamp, I. The effect of the number of aircraft noise events on sleep quality. *Appl. Acoust.* **2014**, *84*, 9–16. [CrossRef]

35. Gasco, L.; Asensio, C.; de Arcas, G. Communicating airport noise emission data to the general public. *Sci. Total Environ.* **2017**, *586*, 836–848. [CrossRef] [PubMed]

36. Dutilleux, G.; Gjestland, T.T.; Licitra, G. Challenges of the Use of Sound Emergence for Setting Legal Noise Limits. *Int. J. Environ. Res. Public Health* **2019**, *16*, 4517. [CrossRef] [PubMed]

37. Bockstael, A.; De Coensel, B.; Lercher, P.; Botteldooren, D. Influence of temporal structure of the sonic environment on annoyance. In *10th International Congress on Noise as a Public Health Problem (ICBEN-2011)*; Griefahn, B., Ed.; Institute of Acoustics: London, UK, 2011; pp. 945–952.

38. Lercher, P.; Bockstael, A.; De Coensel, B.; Dekoninck, L.; Botteldooren, D. The application of a notice-event model to improve classical exposure-annoyance estimation. *J. Acoust. Soc. Am.* **2012**, *131*, 3223. [CrossRef]

39. Lercher, P.; Coensel, B.D.; Dekonink, L.; Botteldooren, D. Community Response to Multiple Sound Sources: Integrating Acoustic and Contextual Approaches in the Analysis. *Int. J. Environ. Res. Public Health* **2017**, *14*, 663. [CrossRef]

40. Lercher, P.; Coensel, B.D.; Dekoninck, L.; Botteldooren, D. Alternative traffic noise indicators and its association with hypertension. In Proceedings of the 11th European Congress and Exposition on Noise Control Engineering (Euronoise 2018), EEA and Greek Acoustical Society Hersonissos, Crete, Greece, 27–31 May 2018; pp. 457–464.

41. Nilsson, M.E.; Botteldooren, D.; Coensel, B.D. Acoustic indicators of soundscape quality and noise annoyance in outdoor urban areas (invited paper). In *19th International Congress on Acoustics*; Instituto de Acústica: Madrid, Spain, 2007.

42. Lercher, P.; Pieren, R.; Wunderli, J.M. Noise and hypertension: Testing alternative acoustic indicators. In Proceedings of the 23rd International Congress on Acoustics, ICA 2019, Aachen, Germany, 9–13 September 2019.

43. Héritier, H.; Vienneau, D.; Foraster, M.; Eze, I.C.; Schaffner, E.; Thiesse, L.; Ruzdik, F.; Habermacher, M.; Köpfli, M.; Pieren, R.; et al. Diurnal variability of transportation noise exposure and cardiovascular mortality: A nationwide cohort study from Switzerland. *Int. J. Hyg. Environ. Health* **2018**, *221*, 556–563. [CrossRef] [PubMed]

44. Brink, M.; Schäffer, B.; Vienneau, D.; Foraster, M.; Pieren, R.; Eze, I.C.; Cajochen, C.; Probst-Hensch, N.; Röösli, M.; Wunderli, J. A survey on exposure-response relationships for road, rail, and aircraft noise annoyance: Differences between continuous and intermittent noise. *Environ. Int.* **2019**, *125*, 277–290. [CrossRef] [PubMed]

45. Wunderli, J.M.; Pieren, R.; Habermacher, M.; Vienneau, D.; Cajochen, C.; Probst-Hensch, N.; Röösli, M.; Brink, M. Intermittency ratio: A metric reflecting short-term temporal variations of transportation noise exposure. *J. Expo. Sci. Environ. Epidemiol.* **2016**, *26*, 575. [CrossRef] [PubMed]

46. Salamon, J.; Bello, J.P. Unsupervised feature learning for urban sound classification. In Proceedings of the 2015 IEEE International Conference on Acoustics, Speech and Signal Processing (ICASSP), Brisbane, QLD, Australia, 19–24 April 2015; pp. 171–175.

47. Socoro, J.C.; Alías, F.; Pages, R.M.A. An Anomalous Noise Events Detector for Dynamic Road Traffic Noise Mapping in Real-Life Urban and Suburban Environments. *Sensors* **2017**, *17*, 2323. [CrossRef] [PubMed]

48. Gloaguen, J.; Can, A.; Lagrange, M.; Petiot, J. Road traffic sound level estimation from realistic urban sound mixtures by Non-negative Matrix Factorization. *Appl. Acoust.* **2019**, *143*, 229–238. [CrossRef]

49. ISO. *ISO/TS 12913-2:2018 Acoustics—Soundscape—Part 2: Data Collection and Reporting Requirements*; ISO: Geneva, Switzerland, 2018.

50. Mitchell, A.; Oberman, T.; Aletta, F.; Erfanian, M.; Kachlicka, M.; Lionello, M.; Kang, J. The Soundscape Indices (SSID) Protocol: A Method for Urban Soundscape Surveys—Questionnaires with Acoustical and Contextual Information. *Appl. Sci.* **2020**, *10*, 2397. [CrossRef]

International Journal of
Environmental Research and Public Health

Article

Effects of Exposure to Road, Railway, Airport and Recreational Noise on Blood Pressure and Hypertension

Davide Petri [1], Gaetano Licitra [2,*], Maria Angela Vigotti [1,3] and Luca Fredianelli [4,*]

[1] Department of Biology, University of Pisa, 56126 Pisa, Italy; davide.petri@unipi.it (D.P.); mariangelavigotti@gmail.com (M.A.V.)

[2] Institute for Chemical-Physical Processes, National Research Council, 56124 Pisa, Italy

[3] Clinical Physiology Institute, National Research Council, 56124 Pisa, Italy

[4] IPool S.r.l., Via Cocchi 7, 56121 Pisa, Italy

* Correspondence: g.licitra@arpat.toscana.it (G.L.); luca.fredianelli@gmail.com (L.F.)

Abstract: Noise is one of the most diffused environmental stressors affecting modern life. As such, the scientific community is committed to studying the main emission and transmission mechanisms aiming at reducing citizens' exposure, but is also actively studying the effects that noise has on health. However, scientific literature lacks data on multiple sources of noise and cardiovascular outcomes. The present cross-sectional study aims to evaluate the impact that different types of noise source (road, railway, airport and recreational) in an urban context have on blood pressure variations and hypertension. 517 citizens of Pisa, Italy, were subjected to a structured questionnaire and five measures of blood pressure in one day. Participants were living in the same building for at least 5 years, were aged from 37 to 72 years old and were exposed to one or more noise sources among air traffic, road traffic, railway and recreational noise. Logistic and multivariate linear regression models have been applied in order to assess the association between exposures and health outcomes. The analyses showed that prevalence of high levels of diastolic blood pressure (DBP) is consistent with an increase of 5 dB (A) of night-time noise (β = 0.50 95% CI: 0.18–0.81). Furthermore, increased DBP is also positively associated with more noise sensitive subjects, older than 65 years old, without domestic noise protection, or who never close windows. Among the various noise sources, railway noise was found to be the most associated with DBP (β = 0.68; 95% CI: −1.36, 2.72). The obtained relation between DBP and night-time noise levels reinforces current knowledge.

Keywords: noise; hypertension; environmental noise; railway noise; recreational noise; airport noise; road traffic noise; blood pressure; noise annoyance; diastolic blood pressure

Citation: Petri, D.; Licitra, G.; Vigotti, M.A.; Fredianelli, L. Effects of Exposure to Road, Railway, Airport and Recreational Noise on Blood Pressure and Hypertension. *Int. J. Environ. Res. Public Health* **2021**, *18*, 9145. https://doi.org/10.3390/ijerph18179145

Academic Editor: Yu-Pin Lin

Received: 3 July 2021
Accepted: 25 August 2021
Published: 30 August 2021

Publisher's Note: MDPI stays neutral with regard to jurisdictional claims in published maps and institutional affiliations.

1. Introduction

Noise pollution represents a great public concern. Long-term exposure to high noise levels (>85 dB) have been associated with many direct health effects, even leading to hearing loss [1,2], or to non-hearing effects when exposure is at low-medium levels [3,4]. In this case, transportation noise can induce annoyance [5–8], sleep disturbance with awakening [9,10], cognitive impairment [11–13], physiological stress reactions [14], endocrine imbalance and cardiovascular disorders [15–18]. Moreover, exposure to noise can reduce both workers' and students' performance [19–21]. Higher levels of stress among subjects exposed to noise level higher than 55 dB (A) and increased occurrence of cardiovascular diseases associated with noise level greater than 65 dB (A) have also been reported [22]. Most of all, hypertension is the leading risk factor for cardiovascular morbidity and mortality worldwide [23]. Indeed, hypertension is a major risk factor for premature death and disability from heart disease, stroke, peripheral vascular disease and kidney failure [24–26]. A meta-analysis [24] evidenced a significant rise in prevalence of hypertension per increase of 5 dB (A) of equivalent road traffic noise level A weighted over a 16 h period ($L_{Aeq,16h}$) (Odds Ratio (OR) = 1.03; 95% confidence interval (CI): 1.01–1.06). Moreover, results on the association between long-term exposure to noise and blood pressure (BP) are still heterogeneous [27–30]. A possible explanation was provided by Babisch [31–33], who suggested that an increase in the level of adrenaline, noradrenaline and cortisol in response to noise-induced stress could result in peripheral vasoconstriction, increased heart rate and a rise in arterial blood pressure. A lack of data on multiple sources of noise and cardiovascular outcomes is still an issue in the scientific literature.

Health impact assessment studies estimated that 104 million U.S. citizens have sufficient annual noise exposure to be at risk of noise-related health effects [34]. In Europe, even 15 years after the implementation of the Environmental Noise Directive [35], 40% of the European population remains exposed to road traffic noise levels over 55 dB (A) of L_{den} (average noise level over a 24 h period) and 15% to levels greater than 65 dB (A). Road traffic remains the most widespread source in the urban environment, followed by railway noise, with 22 million people exposed to noise levels higher than 55 dB (A) of L_{den} [36], then by aircraft noise with more than 4 million people, and industrial noise with 1 million people exposed. The scientific community has studied how different sources generate noise and how to mitigate this with innovative solutions, especially in an urban context with its main sources being road traffic [37,38], railway traffic [39], airport [40,41], industries [42,43] and port activities [44,45], where present.

The impact of air traffic noise is particularly relevant during take-off and landing phases [46,47] if ground taxing operations are incorrectly managed [48]. Specific studies have been dedicated to aircraft noise's relation with sleep disturbances and annoyance [49–51], while others, including the HYENA and SERA projects, focused on blood pressure and the risk of hypertension [46,52–54]. The HYENA project aimed at assessing the impacts on cardiovascular health of noise generated by air traffic and road traffic near six European airports. The results showed significant exposure–response association between night-time aircraft noise, daily road traffic noise and prevalence of "heart disease and stroke" and hypertension

Railways received specific attention in the ALPNAP study [55,56], where significant associations between railway noise and sleep medication intake were shown, especially for people exposed to 60 dB L_{den}. While the study was performed in an Alpine valley characterized by very specific noise conditions, other authors studied the association of railway noise with sleep disturbance [4,57]. Furthermore, railway noise is often related to vibrations, which induce other negative effects on sleep [58,59]. In a previous study [60], the authors showed that railway noise maps underestimate noise exposure and people are disturbed by unconventional noises such as brakes, squeals, whistles, and screeches, which are usually not considered in noise modelling. The underestimation of noise and the presence of vibration resulted in an increase of the percentage of highly annoyed people (%HA) with respect to the traditional noise dose–effect curves.

In an urban context, recreational noise plays an important role in citizens' disturbance, even if it has not yet been well studied yet. In recent years more attention has been paid to the topic [61], but most of studies have only focused on campus students [62]. While the relation of recreational noise to cardiovascular outcomes still needs study, the insurgence of tinnitus, hearing loss and noise-induced hearing-threshold shift due to high levels of music were investigated [63,64] and connections were found by different authors [65].

The present study aims to evaluate the impact that different noise sources have on the health of citizens in terms of blood pressure (BP) and hypertension. A sample of 517 citizens living in the city of Pisa, Italy, was chosen for blood pressure measurements and a structured questionnaire. The city of Pisa was a good test site for the study because of its complex structure in terms of noise sources, including all the previously mentioned transportation sources, with an important airport very close to the inhabited areas, and major roads and railway stretches crossing the residential area. The exposure to all of these sound sources was considered as a whole or individually in order to evaluate their eventual correlation with health parameters. In a public health context, the results obtained could be used by institutions and citizens to prevent exposure to specific noise sources.

2. Materials and Methods

2.1. Sample Selection

The study sample includes 517 subjects, 37–72 years of age at the time of interview, previously selected for the SERA project (study on the effects of airport noise) [66] and the SERA-FA project (study on the effects of airport, railway and recreational noise) [67]. For both projects, the population sample was recruited through a random selection, stratified by gender, age and main sound source from the database of addresses provided by the local General Registry Office. The subjects were extracted uniformly considering sex, age and potential exposure to the principal noise sources according to the noise map of the city. Subsequently, up to three substitutions were selected in order to replace the non-respondents and those who refused to participate.

In the SERA project, the population was recruited in 2012 in a cross-sectional study, with a random sample of adults (45–70 years of age) living in Pisa and exposed to different average noise levels. A first set was exposed to at least 55 dB (A) of airport L_{den}, a second was exposed to 50 dB (A) of both airport and traffic L_{den}, a third was exposed to at least 55 dB (A) of traffic L_{den} and the last was not exposed to significant noise levels from these main sources. Participants were subjected to blood pressure measurements and to a structured questionnaire using the model adopted in the HYENA study [52]. This included questions on house characteristics, possible protection from noise, windows, socio-demographic conditions, occupational noise exposure, dietary habits, lifestyle factors, smoking, noise annoyance, sleep conditions and noise sensitivity.

From 2014 to 2016 more participants, aged 37–72, were added to the SERA-FA study in order to include subjects exposed to at least 55 dB(A) of railway L_{den} and subjects exposed to at least 55 dB (A) of recreational noise, in terms of L_{night}. The same protocol as the SERA project was used, but two sections were added to the questionnaire in order to specifically investigate the exposure to railway and recreational noise.

The questionnaire campaign with the assessment of BP was carried out in 2012–2013 for the SERA project and 2014–2015 for SERA-FA participants.

2.2. Exposure Assessment

Noise exposure to the transport infrastructure (road, railway, airport) was obtained by the noise maps developed by the Environment Protection Agency for the Tuscany Region (ARPAT) according to the guidelines of the European Noise Directive 2002/49/EC (END) and the Italian Decree of 2005 (D. Lgs 194/05) [68]. Using the proper input data required by the noise model (i.e., traffic flow, speed), annual average L_{den} and L_{night} of the single source were computed on a grid of 5 m × 5 m positioned at a height of 4 m above the ground, at a distance of 1 m from the building's façade using the Integrated Noise Model 7.0b (INM) [69]. The overall noise exposure was also calculated as the energetic sum of the three components. These were used to estimate the percentage of residents exposed to noise levels greater than 55 dB (A) L_{den} and 50 dB (A) L_{night}. L_{den} and L_{night} were calculated. The German national method VBEB [70] was used as methodology to assign population to noise levels, as a study reported [71] how this better describes real exposure for epidemiological studies, with respect to the method proposed by the END. VBEB distributes the population among the receiver points located around buildings equally, and determines an exposure proportional to noise levels along all the building's façades, while the END assigns the maximum level from all the points around the corresponding building, which is usually on the most exposed façade [72]. The meteorological parameters considered in the model, such as air temperature, atmospheric pressure, wind speed and relative humidity, were measured by weather stations located in the city of Pisa. Moreover, a measurement campaign for railway noise was conducted in two different parts. In 2013–2014 [73], measurements were performed along the railway lines in 31 places within the city of Pisa with the aim of validating the railway noise map. A class 1 sound level meter, compliant with IEC 61672-1 [74], was placed at a height of 1.5 m and 1 m away from the most exposed façade, recording the A-weighted equivalent continuous sound level (L_{Aeq}) with a time step equal to one second. From February to April 2015, the number of measurement points was increased, with another 27 short term and seven measurements providing daily and nightly value for noise exposure. A comparison between noise measurements and the noise map showed that railway infrastructure affects the surrounding areas differently than forecasted, due to the presence of unconventional noise from maneuvering, loading and unloading, truck movements, braking, squeals, whistles, arrivals and departures of trains, speakers, passengers, internal work, generators, bells, crossings, etc. [60]. The resulting differences have been used to correct the citizens' exposure to railway noise.

At present, no model can simulate recreational noise, thus a specific measurement campaign which lasted for 18 months was conducted in order to assess the areas within the city of Pisa more subject to this source, such as the city center. Noise data were acquired with the wireless sensor network for real-time noise mapping used in the SENSEable project [75]. L_{Aeq} was acquired simultaneously in six different positions with a temporal base equal to one second, averaged in day-time periods. The measurement points were selected [76] based on the number of residents, in order to optimize the search for similar environments from an acoustic point of view. These are the largest areas possible in which it is possible to assume that the sound pressure level varies within 5 dB (A). The monthly average L_{Aeq} was calculated, eliminating occasional sound events, rain and wind. Recreational noise was defined as that part of noise that exceeded the road traffic noise level resulting from the noise map of the area, as this is the only other noise source affecting the city center. Further details on elaboration and stability can be found in the literature [77]. Estimates were then calculated using the main European indicators (L_{den}, L_{night}), with standard deviation as a measure of uncertainty, and were assigned to citizens living in similar environments from an acoustic point of view.

Geographical coordinates were assigned to subjects using a common GIS software. For the addresses geocoding, the normalization and georeferencing service of the Tuscany Region has been used. Residents were classified depending on the superposition of noise maps (L_{den} < 55, 55–59, 60–64, 65–69, ≥70).

2.3. Assessment of the Outcome

Trained interviewers measured systolic blood pressure (SBP) and diastolic blood pressure (DBP) at subject's homes after at least five minutes of rest in a seated position keeping both feet on the ground, using an automatic Omron M6 Comfort model (OMRON, Tokyo, Japan) with cuff attached to right or left upper arm (preferably right) [78]. The visits were performed during day-time from Monday to Friday. The staff assessed SBP and DBP three times at each home visit, with the first measurement recorded at the beginning of the interview after 5 min rest, and the second after a further minute, in according with recommendations of the American Heart Association [79]. The third measurement was at the end of the interview, approximately 45 min later. Home visits were distributed over the day in order to account for possible diurnal variations in BP. Two additional measurements were self-made by subjects in the evening of the same day and in the morning of the following day. The average of the 5 measurements provided the SBP and DBP values used in the analysis.

2.4. Covariates

An evaluation of the possible major confounders was performed among the variables which can be risk factors for hypertension and possibly associated with noise exposure, in order to eventually exclude them from the model. The potential confounders or effect modifiers that we have evaluated were the usual health indicators (physical activity and body mass index—BMI), sociodemographic characteristics (sex, age, education and employment status), lifestyle habits (smoking and alcohol), other noise sources different than mean noise exposure, work-related noise exposure, noise sensitivity value based on standardized ten questions [80,81] and home conditions (double-glass windows, other noise protections, construction year of the house). Subjects also indicated their annoyance to noise on a 11-point scale for each source on a list of ten: this parameter was evaluated as a potential effect modifier of the investigated relationship.

2.5. Statistical Analyses

Standard statistical methods were applied using STATA 14.2 [82]. In addition to SBP and DBP, the prevalence of hypertension based on the self-reported diagnosis was calculated, together with the use of antihypertensive medication or blood pressure measurements reporting SBP $\geq$ 140 mmHg and DBP $\geq$ 90 mmHg. This criterion is recommended by the World Health Organization (WHO) [83].

Pair-wise correlation between noise map indicators (airport, traffic and railway) were calculated and the association between noise levels and hypertension investigated using a logistic regression model. The odds ratio (OR) and 95% Confidence Intervals (CIs) for each effect estimate were estimated as results of this analysis.

The possible relation between environmental noise levels and BP, expressed as SBP and DBP separately, was assessed with mixed linear regression models and associations expressed with both day-time and night-time noise levels, obtaining risk beta coefficients and 95% CIs.

The analysis in categories made of intervals equal to 5 dB (A) suggested a linear relation, thus continuous exposure data have been used to assess the effect estimate in order to increase statistical power.

Potential covariates were evaluated in non-adjusted analysis: those with a $p < 0.20$ in order to avoid exclusion of important adjustment variables due to stochastic variability [84,85] and those already known in literature as risk factors for hypertension [86] (sex, age, BMI, educational status) were selected. The final model included sex, age (as continuous), educational status (elementary, medium, high school, university), alcohol (never drinker, former drinker and actual drinker), physical activity (less than 1 time a week of moderate exercise, between 1 and 3 times a week, more than 3 times a week), BMI, and use of pre-cooked foods (at least once a week, less than 1 meal at week). Smoking was included only in the model for BP, as a well-known risk factor for heart disease, but not for hypertension, as confirmed by the p-value, therefore not relevant in the preliminary analysis.

In order to investigate the differential susceptibility to noise exposure in subgroups of the study population, a stratification of the analysis was performed by sex, age, noise sensitivity (<50th percentile (P50) vs. $\geq$50th percentile), house noise protection (yes vs. no), windows closed to prevent noise exposure (never, few vs. often, always), living room exposition (noise source vs. side of noise source vs. back of noise source), bedroom exposition (noise source vs. side of noise source vs. back of noise source) and annoyance (few, moderately annoyed vs. very annoyed).

3. Results

A total of 517 participants (228 men and 289 women), aged between 35 and 72 years, at the time of visit, participated to the present study. The response rate in the study was medium-low (29.1%). In order to assess the potential selection bias, the authors compared the source population and the sample by sex and age, finding no statistically significant difference.

The mean age of participants was 57.3 years old (standard deviation 8.7) and 44.1% were males. Mean SBP and DBP expressed in mmHg during the visit were 126.9 and 81.1, respectively, while means for self-measured blood pressure were 125.6/79.2 and 121.5/77.7 respectively, for evening and next day morning. The overall hypertension prevalence was 37.5% (to be compared with the Italian population, in which there is a prevalence of 33% and 31% respectively among males and females [87]); of all subjects, 20.1% were treated for hypertension and had normal values of BP, 11.0% were treated but presented hypertensive values of SBP or DBP, and 11.2% without a medical prescription for hypertension presented abnormal values of BP. The prevalence of hypertension was higher in males (44.6%) than in females (32.7%).

Table 1 describes the variables considered in the study and stratified by hypertensive condition expressed as the WHO classification. Statistically significant differences between the two groups arose. Among those with hypertensive condition, higher values of SBP and DBP, BMI, alcohol consumption, lower level of education, actual workers, less than one time/week of moderate exercise, use of precooked foods and lower attitude to close windows were found.

Table 1. Characteristics of subjects included in the study, variables divided in continuous and categorical variables.

Characteristics	Total (n = 515)	Non-Hypertensive (n = 313)	Hypertensive (n = 194)	*p*-Value [a]
Continuous variables [median (IQR)]				
Systolic blood pressure (mmHg)	123.1 (20.0)	117.5 (16.0)	136.5 (20.5)	<0.001 *
Diastolic blood pressure (mmHg)	78.5 (12.1)	76.0 (9.25)	86.8 (13.25)	<0.001 *
Age (years)	58.2 (14.2)	54.8 (12.8)	62.7 (11.3)	<0.001 *
Main noise source L_{DEN} [dB(A)]	62 (10.0)	61.6 (10.6)	62.5 (10.2)	0.066
Main noise source L_{NIGHT} [dB(A)]	53.5 (18.0)	53.1 (15.4)	54.1 (18.3)	0.254
Noise sensitivity score (10–60) [b]	39.0 (12.0)	39.0 (12.0)	39.0 (11.0)	0.874
Categorical variables [n (%)]				
Male sex	228 (44.1)	124 (39.6)	101 (50.2)	0.042
BMI (Kg/m^2)				<0.001 *
<18.5	9 (1.7)	8 (2.6)	1 (0.5)	
18.5–24.9	251 (48.6)	178 (56.9)	73 (36.3)	
25–29.9	201 (38.9)	97 (31.0)	103 (51.2)	
30+	56 (10.8)	30 (9.6)	24 (11.9)	
Educational level				<0.001 *
University or similar	241 (46.6)	166 (53.0)	73 (36.3)	
Secondary	174 (33.7)	103 (32.9)	70 (34.8)	
Primary	66 (12.8)	31 (9.9)	35 (17.4)	
Illiterate	33 (6.4)	12 (3.8)	21 (10.5)	
Smoking				0.501
Professional Status				
Unemployed	58 (11.5)	34 (10.9)	24 (12.5)	
Retired	157 (31.1)	78 (24.9)	79 (24.9)	
Actual worker	290 (64.6)	201 (64.2)	89 (64.2)	<0.001 *
Physical activity (moderate exercise)				
Less than 1 time/week	45 (8.9)	20 (6.4)	25 (13.0)	
Between 1 and 3 times/week	134 (26.5)	91 (29.2)	43 (22.4)	
More than 3 times/week	326 (64.6)	202 (64.5)	124 (64.6)	0.020
Never smokers	235 (45.5)	148 (47.3)	85 (42.3)	
Smokers	116 (22.4)	72 (23.0)	44 (21.9)	
Former smokers	166 (32.1)	93 (29.7)	72 (35.8)	
Drinking [c]				0.009 *
Non-drinkers	146 (28.2)	98 (31.3)	48 (23.9)	
Casual drinkers	187 (38.2)	117 (37.4)	68 (33.8)	
Regular drinkers	183 (35.4)	98 (31.3)	84 (41.8)	
Diet, use of pre-cooked foods	208 (40.2)	142 (45.4)	64 (31.8)	0.009 *
Living room orientation, noise source [d]	128 (24.8)	77 (24.6)	51 (25.4)	0.882
Bedroom orientation, noise source [d]	134 (25.9)	84 (26.8)	48 (23.9)	0.571
Closing windows, yes [e]	162 (31.3)	107 (34.2)	54 (26.9)	0.081
Protections, yes [f]	333 (64.4)	207 (66.1)	124 (61.7)	0.395
Noise annoyance [g]				0.878
Moderate (0–7)	201 (39.0)	119 (45.9)	75 (45.2)	
High (8–10)	231 (44.9)	140 (54.1)	91 (54.8)	
Not exposed	83 (16.1)	54 (67.5)	26 (32.5)	
Air pollution annoyance, high [h]	281 (54.4)	170 (54.3)	110 (54.7)	0.200

Table 1. *Cont.*

Characteristics	Total (n = 515)	Non-Hypertensive (n = 313)	Hypertensive (n = 194)	*p*-Value [a]
Noise groups				0.018
Aircraft	100 (19.8)	59 (18.9)	41 (21.4)	
Road traffic + Aircraft	80 (15.8)	40 (12.8)	40 (20.8)	
Road traffic	74 (14.7)	42 (13.4)	32 (16.7)	
Railway	118 (23.4)	85 (27.2)	33 (17.2)	
Recreational	53 (10.5)	33 (10.5)	20 (10.4)	
Reference group	80 (15.8)	54 (17.3)	26 (13.5)	

[a] Chi-square test and Kruskall-Wallis test for strata of hypertension with categorical or continuous variables, respectively. [b] Higher noise sensitivity with higher values. [c] Casual: less than 2 glasses/week. (1 missing observation) [d] Control group not included. [e] Yes: always close windows (vs. no: never, only on summer or winter season). [f] Sound-proofed windows or changes in structure due to noise. [g] Referred to the main noise source; in case of aircraft/traffic group, the higher annoyance is selected. [h] Score from 6 to 10 in a 0–10 scale. * Category with a significant association with hypertension.

The mean noise levels of the main noise source considered in this study are 61.7 dB (A) (standard deviation 7.6) of L_{den} and 49.4 dB (A) (standard deviation 13.6) of L_{night}.

Multiple associations between covariates and prevalence of hypertension are shown in Table 2. The results represent the relationships between single parameters and risk of hypertension, net of all the other covariates included concurrently, without the main exposure of noise. Variables such as sex (male), higher age, smoking, higher BMI showed significant positive associations with a higher risk of hypertension. Educational level, stability of work conditions and physical activity showed a protective effect, in a significant association with a lower risk of hypertension.

Table 2. Multiple associations between covariates and the prevalence of hypertension (HYENA definition).

Variable	Categorization	OR [a] (95% CI)	*p*-Value
Age	Per 1 year	1.08 (1.05–1.10)	<0.001
Gender	Male	1	
	Women	0.70 (0.46–1.08)	0.107
Alcohol	Never drinker	1	
	Casual drinker	1.25 (0.75–2.10)	0.394
	Regular drinker	1.30 (0.76–2.20)	0.338
Smoking	Never smoker	1	
	Former smoker	0.79 (0.49–1.26)	0.321
	Actual smoker	0.69 (0.41–1.17)	0.169
Professional status	Unemployed	1	
	Retired	0.38 (0.19–0.75)	0.006
	Actual worker	0.49 (0.26–0.92)	0.027
BMI	Per kg/m^2	1.07 (1.02–1.13)	0.010
Noise sensitivity	Per scale unit	1.01 (0.99–1.03)	0.225
Educational level	Illiterate	1	
	Primary	0.91 (0.36–2.32)	0.844
	Secondary	0.61 (0.26–1.41)	0.284
	University or similar	0.47 (0.20–1.09)	0.117
Physical activity	None	1	
	Moderately or strenuous 1–3 times a week	0.43 (0.20–0.91)	0.028
	Moderately or strenuous > 3 times a week	0.57 (0.28–1.13)	0.108

[a] Odds Ratio are mutually adjusted.

Table 3 shows correlation coefficients between environmental noise exposure levels by day and night. Values display a different correlation for each noise source (r = 0.22 for airport noise, 0.99 for both railway and traffic exposures). In addition, significant correlation values between airport noise and railway noise during nighttime were detected, whilst for railways, this seems to be at the boundary of significance during daytime.

Table 3. Descriptive statistics and results of Spearman's correlation of the different environmental noise exposures.

	Mean ± SE	Percentile			Air Traffic		Railway		Road Traffic	
		10th	50th	90th	Day	Night	Day	Night	Day	Night
Air Traffic (day)	57.00 ± 0.20	54.5	57.4	59.3	1					
Air traffic (night)	27.78 ± 0.87	20	25.6	41.1	0.22	1				
Railway (day)	59.53 ± 0.78	46.2	61	70.3	−0.12	0.24	1			
Railway (night)	52.49 ± 0.78	39.2	54.3	63.1	−0.12	0.25	0.99 *	1		
Traffic (day)	68.04 ± 0.36	63.9	68	72	0.05	0.00	−0.07	−0.07	1	
Traffic (night)	59.15 ± 0.37	55.3	59.2	63.4	0.05	0.02	−0.06	−0.06	0.99 *	1
Recreational (day) [a]	70.03 ± 0.66	64.2	71.2	74.2						
Recreational (night) [a]	63.80 ± 0.72	57.5	65.6	68.4						

[a] Recreational noise information are missing for the other types of noise. * Significant value.

The regression model shown in Table 4, indicates that a 5 dB (A) increase in nocturnal environmental noise corresponds to a significant increase in blood pressure, especially in DBP (DBP and night-time noise: β = 0.50, 95% confidence intervals (CIs): 0.18, 0.81). Considering the hypertensive outcome, associations are almost significant especially during night-time in the full adjusted model (OR = 1.07, 95% CI: 0.99–1.15). Night-time noise is involved too in the association with SBP, showing a nearly significant association (β = 0.47, 95% CI: −0.05, 1.00).

Table 4. Associations between hypertension and blood pressure with environmental noise by day and night; estimated risk for hypertension and change in blood pressure (mmHg) for a 10 dB (A) increment during the day or for a 5 dB (A) increment during night.

Outcome		Night		Day	
		OR/5 dB(A) (95% CI)	*p*-Value	OR/10 dB(A) (95% CI)	*p*-Value
Hypertension	Non-adjusted	1.03 (0.96, 1.10)	0.386	1.23 (0.97, 1.57)	0.091
	Full model	1.07 (0.99, 1.15)	0.070	1.27 (0.97, 1.67)	0.085
		β/5 dB (A) (95% CI)		β/10 dB (A) (95% CI)	
SBP	Non-adjusted	0.11 (−0.47, 0.69)	0.715	−0.31 (−2.38, 1.76)	0.768
	Full model	0.47 (−0.05, 1.00)	0.078	−0.08 (−1.97, 1.81)	0.934
DBP	Non-adjusted	0.28 (−0.05, 0.61)	0.101	0.30 (−0.88, 1.48)	0.615
	Full model	0.50 (0.18, 0.81)	0.002	0.91 (−0.23, 2.06)	0.118

Stratifying the main characteristics, the effects estimates were higher in participants who showed a higher noise sensitivity (based on Weinstein's noise sensitivity method). Association between hypertension and environmental nocturnal noise were found in males (OR = 1.13; 95% CI: 1.01, 1.26), in persons older than 65 years of age (OR = 1.18; 95% CI: 1.02, 1.37) and those with higher noise sensitivity (OR = 1.12; 95% CI: 1.01–1.24).

Relations between DBP and environmental nocturnal noise showed some significant results too, among all participants, females (β = 0.40; 95% CI: 0.00, 0.79), people aged over 65 years (β = 1.03; 95% CI: 0.43, 1.62), people moderately annoyed by noise (β = 0.66; 95% CI: 0.05, 1.27) and other categories, shown in Figure 1, as noise sensitivity below and above the 50th percentile (45 in a scale from 10 to 60), structural changes for house noise protection (yes vs. no) and the habit of closing windows (never, few vs. often, always).

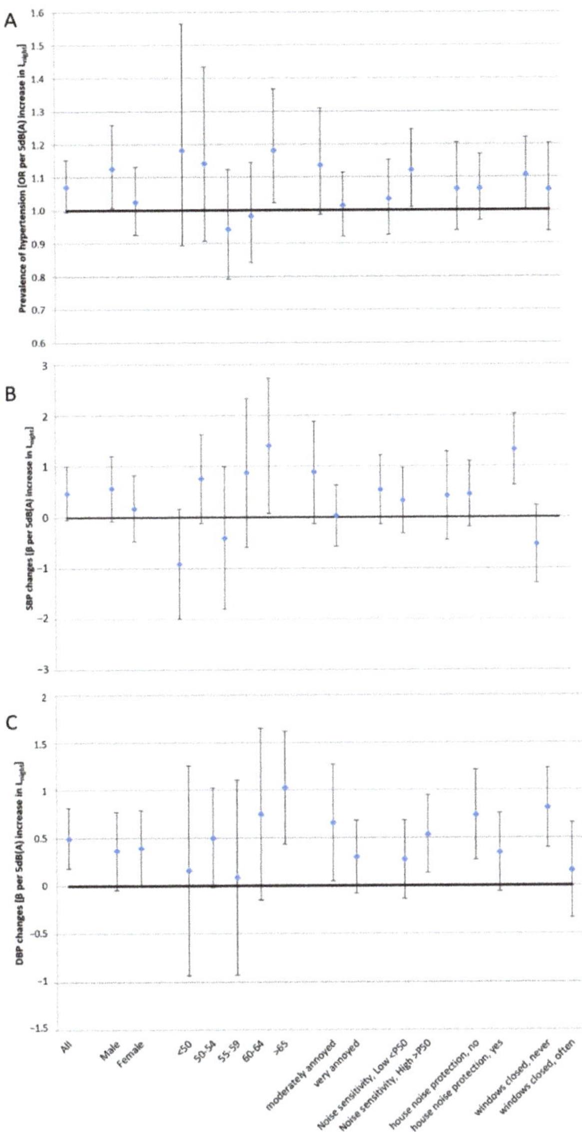

Figure 1. Estimated ORs/β per increment of 5 dB(A) of night-time noise by subgroups of population. in prevalent hypertension (**A**), estimated change of SBP (**B**) and estimated changes of DBP (**C**).

Noise could be quite different in terms of frequency, amplitude and duration of exposure. Figure 2 reports β for an increment of 10 dB(A) in L_{den}, stratified by main noise exposure: only railway exposure has a positive value (β = 0.68; 95% CI: −1.36, 2.72), although it does not reach a statistically significant level of risk.

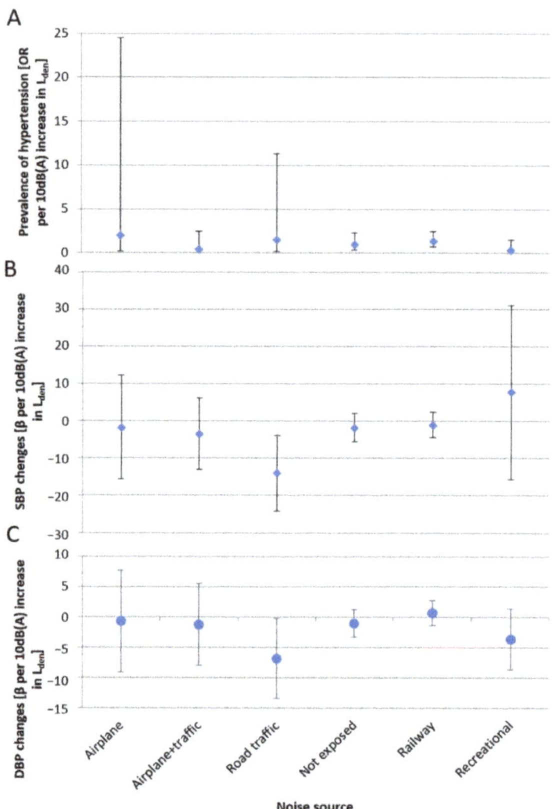

Figure 2. Estimated ORs/β per increment of 5 dB(A) of noise by main noise exposure. in prevalent hypertension (**A**), estimated change of SBP (**B**) and estimated changes of DBP (**C**).

4. Discussion

The present study investigated the impact of exposure to multiple noise sources on the blood pressure and on onset of hypertension in the wake of the HYENA study. Significant positive exposure–response relationships, especially to night-time noise exposure, were found in males; people aged over 65 years old and with a high sensitivity to noise in association with hypertension,; people aged over 65 years old and who never close windows at home due to noise in association with increase in systolic blood pressure; and for all participants, females, people aged over 65 years old, moderately annoyed by noise, high sensitivity to noise, without house noise protection and who never close windows due to noise, in association with an increase in diastolic blood pressure.

Significant differences were found in the hypertensive outcome between the various noise sources, considered to be road, railway and airport traffics and recreational noise. The exposure-response relationships between sound levels and cardiovascular outcomes showed different ORs depending on the sound sources analyzed; railway noise showed highest ORs. It should be noted that railway and road traffic noise were highly correlated between day and night, unlike aircraft noise. A possible explanation is that aircraft activity in Pisa is limited during the night. Nevertheless, the L_{night} indicator (10 p.m.–6 a.m.) includes the so-called "shoulder hours" of the late evening and early morning, where some planes fly in an environment with a background noise lower than that in the day-time.

It is reasonable to believe that the relationship highlighted between nocturnal noise and hypertension can be motivated by the fact that the participants spent the night inside their houses compared to daytime hours, since the noise level assigned to the home address was used. This procedure would also explain the lower misclassification exposure during night hours, compared to daytime. Indeed, it is therefore reasonable to assume that during night hours the participants were actually subjected to the sound levels shown and that therefore there may be a correlation with cardiovascular effects, as emerged from the analyzes [88].

Smoking and alcohol are historical risk factors for hypertension, although smoking is still under investigation for its effect on blood pressure. For this reason, subjects were asked to refrain from smoking during the 30 min before the interview and BP measurement. As detailed in the methods section, smoking was included as a variable in the model, even if its impact on estimate of the exposure–response was not relevant.

The exposure–response association for hypertensive risk was more relevant among men, in accordance with previous evidence on males and hypertension [89–91]. However, the studies mentioned only investigated the relation between road traffic noise and hypertension. The present study aimed to consider a larger number of noise sources in a city like Pisa, where citizens are often exposed to a mixture of noise pollutants. Even when transportation noise seems to be absent, such as in the city center, anthropogenic noise could play a role among the determinants of cardiovascular and sleep health.

In all the investigated outcomes (hypertension, SBP and DBP), variable "age" gave the same indication of higher risk for people aged over 65 years, given that this category is likely to spend more time at home, and consequently its exposure should be less commonly misclassified.

Age always shows positive values and reaches significance in all night-noise analysis for 5 dB(A) increase in L_{night} (Hypertension OR = 1.18; 95% CI: 1.02–1.37; SBP β = 1.41; 95% CI: 0.08–2.73; DBP β = 1.03; 95% CI: 0.43–1.62).

Gathering together some of the subcategories detected in the questionnaire, as reported in Figure 1, it emerged that, in the relation between DPB increase and L_{night} noise, significant risks were found in subjects moderately annoyed ((0–7 in a 11-point scale) β = 0.66; 95% CI, 0.05–1.27), or with lower noise sensitivity, beyond 50th percentile, (β = 0.54; 95% CI, 0.13–0.94), or living in a house free of noise protection (β = 0.74; 95% CI, 0.26–1.21) or who never close windows because of noise (β = 0.82; 95% CI, 0.40–1.23). Apparently, people who not protecting themselves from noisy sources for personal reasons are more at risk than those who, concerned about the possible effects of noise on their health, strives to protect themselves from this specific exposure.

A potential weakness of this study is the medium-low response rate. However, a descriptive analysis showed that response rate is not different by sex, age and exposure zones, the only exception being the aircraft noise group showing a higher response rate. This can be partially explained by taking into account the limited population of the city (almost 90,000 residents) with a high component of students, and the airport, which is very close to the city, represents the major environmental concern of citizens.

Another weakness of the study could be the different exposure assessment of recreational noise, involving no initial "pedestrian data flow", several microphones in specific areas and a subsequent model. A misclassification and a problem of comparison between noise sources could exist, as no data on recreational noise outside of the city center of Pisa were available. On these bases, the recreational group resulted in a non-significant and negative relation in all analyses, therefore the results, including all areas, could be underestimates.

At the same time, the present study focused not on a single type of noise, but shifted attention towards a more comprehensive approach to noise exposure that involves citizens in several ways, each with its peculiarities (frequency content, amplitude, individual perception, etc.). In addition, the completeness of the questionnaire helped to clarify certain factors and their roles in the associations investigated.

5. Conclusions

Statistically significant positive relation between night-time noise and diastolic blood pressure was found. The subcategories majorly involved in the relation between night-time noise and diastolic blood pressure were people aged older than 65 years, moderately annoyed, noise sensitive, without noise protections in house and residents who usually do not close windows when exposed to high levels of noise. Among various noise sources, railway noise showed the strongest association with the outcomes of the study. Hypertension is a major independent risk factor for events such as myocardial infarction and stroke and this study demonstrated an increase of risk in association with environmental noise.

Author Contributions: Conceptualization, D.P. and M.A.V.; methodology, D.P., G.L., M.A.V. and L.F.; software, D.P. and L.F.; validation, D.P., G.L., M.A.V. and L.F.; formal analysis, D.P. and L.F.; investigation, D.P. and L.F.; resources, D.P., G.L., M.A.V. and L.F.; data curation, D.P. and L.F.; writing—original draft preparation, D.P. and L.F.; writing—review and editing, D.P., G.L., M.A.V. and L.F.; visualization, D.P. and L.F.; supervision, M.A.V.; project administration, M.A.V.; funding acquisition, M.A.V. All authors have read and agreed to the published version of the manuscript.

Funding: This research received no external funding.

Institutional Review Board Statement: The present study protocol was evaluated and accepted by local ethics commission (Comitato Etico Scientifico dell'Azienda Ospedaliera Universitaria di Pisa, Protocollo 3486 del 2012).

Informed Consent Statement: Informed consent was obtained from all subjects involved in the study.

Data Availability Statement: The data presented in this study are available on request from the corresponding author.

Acknowledgments: The research was supported by Italian Ministry of Health. We thank our colleagues of Clinical Physiology Institute in Pisa who assisted our research. The authors wish to thank Lara Ginevra Del Pizzo for providing proof reading.

Conflicts of Interest: The authors declare no conflict of interest.

References

1.	Kujawa, S.G.; Liberman, M.C. Acceleration of age-related hearing loss by early noise exposure: Evidence of a misspent youth. *J. Neurosci.* **2006**. [CrossRef]
2.	Sliwinska-Kowalska, M.; Davis, A. Noise-induced hearing loss. *Noise Health* **2012**. [CrossRef]
3.	Miedema, H.M.E.; Vos, H. Associations between self-reported sleep disturbance and environmental noise based on reanalyses of pooled data from 24 studies. *Behav. Sleep Med.* **2007**. [CrossRef] [PubMed]
4.	Basner, M.; McGuire, S. WHO environmental noise guidelines for the european region: A systematic review on environmental noise and effects on sleep. *Int. J. Environ. Res. Public Health* **2018**, *15*, 519. [CrossRef] [PubMed]
5.	Miedema, H.M.; Oudshoorn, C.G. Annoyance from transportation noise: Relationships with exposure metrics DNL and DENL and their confidence intervals. *Environ. Health Perspect.* **2001**, *109*, 409–416. [CrossRef] [PubMed]
6.	Eze, I.C.; Foraster, M.; Schaffner, E.; Vienneau, D.; Héritier, H.; Pieren, R.; Thiesse, L.; Rudzik, F.; Rothe, T.; Pons, M.; et al. Transportation noise exposure, noise annoyance and respiratory health in adults: A repeated-measures study. *Environ. Int.* **2018**, *121*, 741–750. [CrossRef] [PubMed]
7.	Sung, J.H.; Lee, J.; Jeong, K.S.; Lee, S.; Lee, C.; Jo, M.-W.; Sim, C.S. Influence of Transportation Noise and Noise Sensitivity on Annoyance: A Cross-Sectional Study in South Korea. *Int. J. Environ. Res. Public Health* **2017**, *14*, 322. [CrossRef]
8.	Guski, R.; Schreckenberg, D.; Schuemer, R. WHO environmental noise guidelines for the European region: A systematic review on environmental noise and annoyance. *Int. J. Environ. Res. Public Health* **2017**, *14*, 1539. [CrossRef]
9.	Muzet, A. Environmental noise, sleep and health. *Sleep Med. Rev.* **2007**, *11*, 135–142. [CrossRef] [PubMed]
10.	Park, T.; Kim, M.; Jang, C.; Choung, T.; Sim, K.-A.; Seo, D.; Chang, S.I. The Public Health Impact of Road-Traffic Noise in a Highly-Populated City, Republic of Korea: Annoyance and Sleep Disturbance. *Sustainability* **2018**, *10*, 2947. [CrossRef]
11.	Hygge, S.; Evans, G.W.; Bullinger, M. A prospective study of some effects of aircraft noise on cognitive performance in schoolchildren. *Psychol. Sci.* **2002**, *13*, 469–474. [CrossRef] [PubMed]
12.	Lercher, P.; Evans, G.W.; Meis, M. Ambient Noise and Cognitive Processes among Primary Schoolchildren. *Environ. Behav.* **2003**, *35*, 725–735. [CrossRef]
13.	Clark, C.; Paunovic, K. WHO environmental noise guidelines for the european region: A systematic review on environmental noise and cognition. *Int. J. Environ. Res. Public Health* **2018**, *15*, 285. [CrossRef]

14. Daiber, A.; Kröller-Schön, S.; Frenis, K.; Oelze, M.; Kalinovic, S.; Vujacic-Mirski, K.; Kuntic, M.; Bayo Jimenez, M.T.; Helmstädter, J.; Steven, S.; et al. Environmental noise induces the release of stress hormones and inflammatory signaling molecules leading to oxidative stress and vascular dysfunction—Signatures of the internal exposome. *Biofactors* **2019**, *45*, 495–506. [CrossRef] [PubMed]
15. Basner, M.; Babisch, W.; Davis, A.; Brink, M.; Clark, C.; Janssen, S.; Stansfeld, S. Auditory and non-auditory effects of noise on health. *Lancet* **2014**, *383*, 1325–1332. [CrossRef]
16. Münzel, T.; Gori, T.; Babisch, W.; Basner, M. Cardiovascular effects of environmental noise exposure. *Eur. Heart J.* **2014**, *35*, 829–836. [CrossRef]
17. Dratva, J.; Phuleria, H.C.; Foraster, M.; Gaspoz, J.-M.; Keidel, D.; Künzli, N.; Liu, L.-J.S.; Pons, M.; Zemp, E.; Gerbase, M.W.; et al. Transportation noise and blood pressure in a population-based sample of adults. *Environ. Health Perspect.* **2012**, *120*, 50–55. [CrossRef]
18. Van Kempen, E.; Casas, M.; Pershagen, G.; Foraster, M. WHO environmental noise guidelines for the European region: A systematic review on environmental noise and cardiovascular and metabolic effects: A summary. *Int. J. Environ. Res. Public Health* **2018**, *15*, 379. [CrossRef]
19. Vukić, L.; Mihanović, V.; Fredianelli, L.; Plazibat, V. Seafarers' Perception and Attitudes towards Noise Emission on Board Ships. *Int. J. Environ. Res. Public Health* **2021**, *18*, 6671. [CrossRef]
20. Themann, C.L.; Masterson, E.A. Occupational noise exposure: A review of its effects, epidemiology, and impact with recommendations for reducing its burden. *J. Acoust. Soc. Am.* **2019**, *146*, 3879. [CrossRef] [PubMed]
21. Caviola, S.; Visentin, C.; Borella, E.; Mammarella, I.; Prodi, N. Out of the noise: Effects of sound environment on maths performance in middle-school students. *J. Environ. Psychol.* **2021**, *73*, 101552. [CrossRef]
22. Berglund, B.; Lindvall, T.; Schwela, D.H. New Who Guidelines for Community Noise. *Noise Vib. Worldw.* **2000**, *31*, 24–29. [CrossRef]
23. Lim, S.S.; Vos, T.; Flaxman, A.D.; Danaei, G.; Shibuya, K.; Adair-Rohani, H.; Amann, M.; Anderson, H.R.; Andrews, K.G.; Aryee, M.; et al. A comparative risk assessment of burden of disease and injury attributable to 67 risk factors and risk factor clusters in 21 regions, 1990–2010: A systematic analysis for the Global Burden of Disease Study 2010. *Lancet* **2012**, *380*, 2224–2260. [CrossRef]
24. Van Kempen, E.; Babisch, W. The quantitative relationship between road traffic noise and hypertension: A meta-analysis. *J. Hypertens.* **2012**, *30*, 1075–1086. [CrossRef] [PubMed]
25. Ward, B.W.; Schiller, J.S. Prevalence of multiple chronic conditions among US adults: Estimates from the National Health Interview Survey, 2010. *Prev. Chronic Dis.* **2013**, *10*, E65. [CrossRef]
26. Ward, B.W.; Schiller, J.S.; Goodman, R.A. Multiple chronic conditions among US adults: A 2012 update. *Prev. Chronic Dis.* **2014**, *11*, E62. [CrossRef]
27. Banerjee, D.; Das, P.P.; Fouzdar, A. Urban residential road traffic noise and hypertension: A cross-sectional study of adult population. *J. Urban Health* **2014**, *91*, 1144–1157. [CrossRef]
28. Floud, S.; Blangiardo, M.; Clark, C.; de Hoogh, K.; Babisch, W.; Houthuijs, D.; Swart, W.; Pershagen, G.; Katsouyanni, K.; Velonakis, M.; et al. Exposure to aircraft and road traffic noise and associations with heart disease and stroke in six European countries: A cross-sectional study. *Environ. Health* **2013**, *12*, 89. [CrossRef]
29. Haralabidis, A.S.; Dimakopoulou, K.; Velonaki, V.; Barbaglia, G.; Mussin, M.; Giampaolo, M.; Selander, J.; Pershagen, G.; Dudley, M.-L.; Babisch, W.; et al. Can exposure to noise affect the 24 h blood pressure profile? Results from the HYENA study. *J. Epidemiol. Community Health* **2011**, *65*, 535–541. [CrossRef]
30. Kearney, P.M.; Whelton, M.; Reynolds, K.; Muntner, P.; Whelton, P.K.; He, J. Global burden of hypertension: Analysis of worldwide data. *Lancet* **2005**, *365*, 217–223. [CrossRef]
31. Babisch, W.; Fromme, H.; Beyer, A.; Ising, H. Increased catecholamine levels in urine in subjects exposed to road traffic noise: The role of stress hormones in noise research. *Environ. Int.* **2001**. [CrossRef]
32. Babisch, W. Stress hormones in the research on cardiovascular effects of noise. *Noise Health* **2003**, *5*, 1–11. [PubMed]
33. Babisch, W. Cardiovascular effects of noise. *Noise Health* **2011**, *13*, 201–204. [CrossRef] [PubMed]
34. Hammer, M.S.; Swinburn, T.K.; Neitzel, R.L. Environmental noise pollution in the United States: Developing an effective public health response. *Environ. Health Perspect.* **2014**, *122*, 115–119. [CrossRef] [PubMed]
35. European Commission. *European Commission Report from the Commission to the European Parliament and the Council on the Implementation of the Environmental Noise Directive in Accordance with Article 11 of Directive 2002/49/EC. COM/2017/0151 Final;* European Commission: Brussels, Belgium, 2017.
36. European Environment Agency (EEA). *Noise in Europe 2014;* European Environment Agency: Copenhagen, Denmark, 2014.
37. Morley, D.W.; de Hoogh, K.; Fecht, D.; Fabbri, F.; Bell, M.; Goodman, P.S.; Elliott, P.; Hodgson, S.; Hansell, A.L.; Gulliver, J. International scale implementation of the CNOSSOS-EU road traffic noise prediction model for epidemiological studies. *Environ. Pollut.* **2015**, *206*, 332–341. [CrossRef]
38. Ruiz-Padillo, A.; Ruiz, D.; Torija, A.; Ramos-Ridao, Á. Selection of suitable alternatives to reduce the environmental impact of road traffic noise using a fuzzy multi-criteria decision model. *Environ. Impact Assess. Rev.* **2016**, *61*, 8–18. [CrossRef]
39. Bunn, F.; Zannin, P.H.T. Assessment of railway noise in an urban setting. *Appl. Acoust.* **2016**, *104*, 16–23. [CrossRef]
40. Gagliardi, P.; Fredianelli, L.; Simonetti, D.; Licitra, G. ADS-B system as a useful tool for testing and redrawing noise management strategies at Pisa Airport. *Acta Acust. United Acust.* **2017**, *103*, 543–551. [CrossRef]

41. Iglesias Merchan, C.; Diaz-Balteiro, L.; Soliño, M. Transportation planning and quiet natural areas preservation: Aircraft overflights noise assessment in a National Park. *Transp. Res. Part D Transp. Environ.* **2015**, *41*, 1–12. [CrossRef]
42. Kephalopoulos, S.; Paviotti, M.; Anfosso-Lédée, F.; Van Maercke, D.; Shilton, S.; Jones, N. Advances in the development of common noise assessment methods in Europe: The CNOSSOS-EU framework for strategic environmental noise mapping. *Sci. Total Environ.* **2014**, *482–483*, 400–410. [CrossRef] [PubMed]
43. Morel, J.; Marquis-Favre, C.; Gille, L.-A. Noise annoyance assessment of various urban road vehicle pass-by noises in isolation and combined with industrial noise: A laboratory study. *Appl. Acoust.* **2016**, *101*, 47–57. [CrossRef]
44. Borelli, D. Maritime Airborne Noise: Ships and Harbours. *Int. J. Acoust. Vib.* **2019**, *24*, 631–632. [CrossRef]
45. Fredianelli, L.; Bolognese, M.; Fidecaro, F.; Licitra, G. Classification of Noise Sources for Port Area Noise Mapping. *Environments* **2021**, *8*, 12. [CrossRef]
46. Eriksson, C.; Rosenlund, M.; Pershagen, G.; Hilding, A.; Ostenson, C.-G.; Bluhm, G. Aircraft noise and incidence of hypertension. *Epidemiology* **2007**, *18*, 716–721. [CrossRef]
47. Huss, A.; Spoerri, A.; Egger, M.; Röösli, M. Aircraft noise, air pollution, and mortality from myocardial infarction. *Epidemiology* **2010**, *21*, 829–836. [CrossRef] [PubMed]
48. Licitra, G.; Gagliardi, P.; Fredianelli, L.; Simonetti, D. Noise mitigation action plan of Pisa civil and military airport and its effects on people exposure. *Appl. Acoust.* **2014**, *84*, 25–36. [CrossRef]
49. Basner, M.; Müller, U.; Elmenhorst, E.-M. Single and combined effects of air, road, and rail traffic noise on sleep and recuperation. *Sleep* **2011**, *34*, 11–23. [CrossRef]
50. Fritschi, L.; Brown, L.; Kim, R.; Schwela, D.H.; Kephalopoulos, S. *Burden of Disease from Environmental Noise. Quantification of Healthy Life Years Lost in Europe*; WHO: Geneva, Switzerland, 2011.
51. Hellmuth, T.; Classen, T.; Kim, R.; Kephalopoulos, S. *Methodological Guidance for Estimating the Burden of Disease from Environmental Noise*; WHO: Geneva, Switzerland, 2012.
52. Jarup, L.; Babisch, W.; Houthuijs, D.; Pershagen, G.; Katsouyanni, K.; Cadum, E.; Dudley, M.-L.; Savigny, P.; Seiffert, I.; Swart, W.; et al. Hypertension and exposure to noise near airports: The HYENA study. *Environ. Health Perspect.* **2008**, *116*, 329–333. [CrossRef] [PubMed]
53. Eriksson, C.; Bluhm, G.; Hilding, A.; Ostenson, C.-G.; Pershagen, G. Aircraft noise and incidence of hypertension—Gender specific effects. *Environ. Res.* **2010**, *110*, 764–772. [CrossRef] [PubMed]
54. Rosenlund, M.; Berglind, N.; Pershagen, G.; Järup, L.; Bluhm, G. Increased prevalence of hypertension in a population exposed to aircraft noise. *Occup. Environ. Med.* **2001**, *58*, 769–773. [CrossRef]
55. Lercher, P.; Brink, M.; Rudisser, J.; Van Renterghem, T.; Botteldooren, D.; Baulac, M.; Defrance, J. The effects of railway noise on sleep medication intake: Results from the ALPNAP-study. *Noise Health* **2010**, *12*, 110–119. [CrossRef]
56. Lercher, P.; Botteldooren, D.; Widmann, U.; Uhrner, U.; Kammeringer, E. Cardiovascular effects of environmental noise: Research in Austria. *Noise Health* **2011**, *13*, 234–250. [CrossRef]
57. Miedema, H.M.E.; Vos, H. Exposure response functions for transportation noise. *J. Acoust. Soc. Am.* **1998**, *104*, 3432–3445. [CrossRef] [PubMed]
58. Persson Waye, K.; Bengtsson, J.; Agge, A.; Björkman, M. A descriptive cross-sectional study of annoyance from low frequency noise installations in an urban environment. *Noise Health* **2003**, *5*, 35–46. [PubMed]
59. Smith, M.G.; Croy, I.; Ogren, M.; Persson Waye, K. On the influence of freight trains on humans: A laboratory investigation of the impact of nocturnal low frequency vibration and noise on sleep and heart rate. *PLoS ONE* **2013**, *8*, e55829. [CrossRef] [PubMed]
60. Licitra, G.; Fredianelli, L.; Petri, D.; Vigotti, M.A. Annoyance evaluation due to overall railway noise and vibration in Pisa urban areas. *Sci. Total Environ.* **2016**. [CrossRef]
61. Ottoz, E.; Rizzi, L.; Nastasi, F. Recreational noise: Impact and costs for annoyed residents in Milan and Turin. *Appl. Acoust.* **2018**, *133*, 173–181. [CrossRef]
62. Le Prell, C.G.; Siburt, H.W.; Lobarinas, E.; Griffiths, S.K.; Spankovich, C. No Reliable Association Between Recreational Noise Exposure and Threshold Sensitivity, Distortion Product Otoacoustic Emission Amplitude, or Word-in-Noise Performance in a College Student Population. *Ear Hear.* **2018**, *39*, 1057–1074. [CrossRef]
63. Henderson, E.; Testa, M.A.; Hartnick, C. Prevalence of noise-induced hearing-threshold shifts and hearing loss among US youths. *Pediatrics* **2011**, *127*, e39–e46. [CrossRef]
64. Serra, M.R.; Biassoni, E.C.; Richter, U.; Minoldo, G.; Franco, G.; Abraham, S.; Carignani, J.A.; Joekes, S.; Yacci, M.R. Recreational noise exposure and its effects on the hearing of adolescents. Part I: An interdisciplinary long-term study. *Int. J. Audiol.* **2005**, *44*, 65–73. [CrossRef]
65. Śliwińska-Kowalska, M.; Zaborowski, K. WHO Environmental Noise Guidelines for the European Region: A Systematic Review on Environmental Noise and Permanent Hearing Loss and Tinnitus. *Int. J. Environ. Res. Public Health* **2017**, *14*, 1139. [CrossRef] [PubMed]
66. Ancona, C.; Golini, M.N.; Mataloni, F.; Camerino, D.; Chiusolo, M.; Licitra, G.; Ottino, M.; Pisani, S.; Cestari, L.; Vigotti, M.A.; et al. Health Impact Assessment of airport noise on people living nearby six Italian airports. *Epidemiol. Prev.* **2014**, *38*, 227–236. [PubMed]
67. Vigotti, M.A.; Petri, D.; Fredianelli, L.; Licitra, G.; Ancona, C. Railway noise and blood pressure levels in Pisa Population, Italy. In Proceedings of the International Symposium of Engineering Education, Manchester, UK, 17–20 June 2014.

68. Panicucci, A. Definizione Della Mappatura Acustica Strategica del Comune di Pisa ai Sensi della Direttiva Europea 49/2002/EC. Ph.D. Thesis, Università di Pisa, Pisa, Italy, 2008.
69. He, H.; Dinges, E.; Hermann, J.; Rickel, D.; Mirsky, L.; Roof, C.; Boeker, E.; Gerbi, P.; Senzig, D.A. *Integrated Noise Model (INM) Version 7.0 User's Guide*; Federal Aviation Administration: Washington, DC, USA, 2007.
70. Umwelt Bundes Amt Vorläufige Berechnungsmethode zur Ermittlung der Belastetenzahlen durch Umge-Bungslärm—VBEB: 2007. Available online: https://www.umweltbundesamt.de/sites/default/files/medien/pdfs/VBEB.pdf (accessed on 29 August 2021).
71. Licitra, G.; Ascari, E.; Brambilla, G. Comparative analysis of methods to evaluate noise exposure and annoyance of people. In Proceedings of the 20th International Congress on Acoustics 2010, ICA 2010—Incorporating Proceedings of the 2010 Annual Conference of the Australian Acoustical Society, Sydney, Australia, 23–27 August 2010.
72. Licitra, G.; Ascari, E.; Fredianelli, L. Prioritizing Process in Action Plans: A Review of Approaches. *Curr. Pollut. Rep.* **2017**, *3*, 151–161. [CrossRef]
73. Fredianelli, L.; Petri, D.; Licitra, G.; Vigotti, M.A. Railways Noise Assessment in Urban Area for Evaluating Health Effect. In Proceedings of the 7th Forum Acusticum, Krakow, Poland, 7–12 September 2014; pp. 1–6.
74. International Electrotechnical Commission. *Electroacoustics—Sound Level Meters—Part 1: Specifications (IEC 61672-1)*; International Electrotechnical Commission: Geneva, Switzerland, 2013.
75. Nencini, L.; De Rosa, P.; Ascari, E.; Vinci, B.; Alexeeva, N. SENSEable Pisa: A wireless sensor network for real-time noise mapping. In Proceedings of the European Conference on Noise Control, Prague, Czech Republic, 10–13 June 2012.
76. Nencini, L.; Vinci, B.; Vigotti, M.A. Setup della rete SENSEABLE Pisa per la realizzazione di uno studio di valuazione degli effetti del rumore antropico sulla salute dei cittadini. In Proceedings of the 41 Convegno Nazionale Associazione Italiana di Acustica, Pisa, Italy, 17–19 June 2014; pp. 17–19.
77. Sotirakopoulos, K.; Barham, R.; Piper, B.; Nencini, L. A statistical method for assessing network stability using the Chow test. *Environ. Sci. Process. Impacts* **2015**. [CrossRef]
78. Ackermann-Liebrich, U.; Kuna-Dibbert, B.; Probst-Hensch, N.M.; Schindler, C.; Dietrich, D.F.; Stutz, E.Z.; Bayer-Oglesby, L.; Baum, F.; Brändli, O.; Brutsche, M.; et al. Follow-up of the Swiss Cohort Study on Air Pollution and Lung Diseases in Adults (SAPALDIA 2) 1991-2003: Methods and characterization of participants. *Soz. Praventivmed.* **2005**. [CrossRef]
79. Pickering, T.G.; Hall, J.E.; Appel, L.J.; Falkner, B.E.; Graves, J.; Hill, M.N.; Jones, D.W.; Kurtz, T.; Sheps, S.G.; Roccella, E.J. Recommendations for blood pressure measurement in humans and experimental animals: Part 1: Blood pressure measurement in humans: A statement for professionals from the Subcommittee of Professional and Public Education of the American Heart Association Co. *Circulation* **2005**, *111*, 697–716. [CrossRef]
80. Luz, G. A tutorial on noise-sensitivity. *J. Acoust. Soc. Am.* **2011**, *129*, 2409. [CrossRef]
81. Weinstein, N.D. Individual differences in reactions to noise: A longitudinal study in a college dormitory. *J. Appl. Psychol.* **1978**, *63*, 458–466. [CrossRef]
82. StataCorp. *Stata Statistical Software: Release 14*; StataCorp LP: College Station, TX, USA, 2015.
83. Chalmers, J.; MacMahon, S.; Mancia, G.; Whitworth, J.; Beilin, L.; Hansson, L.; Neal, B.; Rodgers, A.; Ni Mhurchu, C.; Clark, T. 1999 World Health Organization-International Society of Hypertension Guidelines for the management of hypertension. Guidelines sub-committee of the World Health Organization. *Clin. Exp. Hypertens.* **1999**, *21*, 1009–1060. [CrossRef]
84. Heinze, G.; Dunkler, D. Five myths about variable selection. *Transpl. Int.* **2017**, *30*, 6–10. [CrossRef]
85. Grant, S.W.; Hickey, G.L.; Head, S.J. Statistical primer: Multivariable regression considerations and pitfalls. *Eur. J. Cardio-Thorac. Surg.* **2019**, *55*, 179–185. [CrossRef]
86. World Health Organization. *HEARTS Technical Package*; WHO: Geneva, Switzerland, 2020.
87. Giampaoli, S.; Vescio, M.F.; Gaggioli, A.; Vanuzzo, D. Prevalenza dell'ipertensione arteriosa nella popolazione Italiana. *Bollettino Epidemiol. Naz.* **2002**, *15*, 9.
88. Davies, R.J.; Belt, P.J.; Roberts, S.J.; Ali, N.J.; Stradling, J.R. Arterial blood pressure responses to graded transient arousal from sleep in normal humans. *J. Appl. Physiol.* **1993**, *74*, 1123–1130. [CrossRef] [PubMed]
89. Babisch, W.; Beule, B.; Schust, M.; Kersten, N.; Ising, H. Traffic noise and risk of myocardial infarction. *Epidemiology* **2005**, *16*, 33–40. [CrossRef] [PubMed]
90. Belojevic, G.; Saric-Tanaskovic, M. Prevalence of Arterial Hypertension and Myocardial Infarction in Relation to Subjective Ratings of Traffic Noise Exposure. *Noise Health* **2002**, *4*, 33–37. [PubMed]
91. Herbold, M.; Hense, H.W.; Keil, U. Effects of road traffic noise on prevalence of hypertension in men: Results of the Luebeck Blood Pressure Study. *Soz. Praventivmed.* **1989**, *34*, 19–23. [CrossRef]

International Journal of
Environmental Research and Public Health

Article

Does the Macro-Temporal Pattern of Road Traffic Noise Affect Noise Annoyance and Cognitive Performance?

Beat Schäffer [1,*], Armin Taghipour [1,2], Jean Marc Wunderli [1], Mark Brink [3], Lél Bartha [1,4] and Sabine J. Schlittmeier [5]

1   Empa, Swiss Federal Laboratories for Materials Science and Technology, 8600 Dübendorf, Switzerland; armin.taghipour@hslu.ch (A.T.); jean-marc.wunderli@empa.ch (J.M.W.); lel.bartha@gmail.com (L.B.)
2   School of Engineering and Architecture, Lucerne University of Applied Sciences and Arts, 6048 Horw, Switzerland
3   Federal Office for the Environment (FOEN), 3003 Bern, Switzerland; mark.brink@bafu.admin.ch
4   Faculty of Engineering, University of Regina, Regina, SK S4S 0A2, Canada
5   Institute of Psychology, RWTH Aachen University, 52056 Aachen, Germany; sabine.schlittmeier@psych.rwth-aachen.de
*   Correspondence: beat.schaeffer@empa.ch; Tel.: +41-58-765-47-37

Citation: Schäffer, B.; Taghipour, A.; Wunderli, J.M.; Brink, M.; Bartha, L.; Schlittmeier, S.J. Does the Macro-Temporal Pattern of Road Traffic Noise Affect Noise Annoyance and Cognitive Performance? *Int. J. Environ. Res. Public Health* **2022**, *19*, 4255. https://doi.org/10.3390/ijerph19074255

Academic Editor: Paul B. Tchounwou

Received: 11 February 2022
Accepted: 28 March 2022
Published: 2 April 2022

Publisher's Note: MDPI stays neutral with regard to jurisdictional claims in published maps and institutional affiliations.

Abstract: Noise annoyance is usually estimated based on time-averaged noise metrics. However, such metrics ignore other potentially important acoustic characteristics, in particular the macro-temporal pattern of sounds as constituted by quiet periods (noise breaks). Little is known to date about its effect on noise annoyance and cognitive performance, e.g., during work. This study investigated how the macro-temporal pattern of road traffic noise affects short-term noise annoyance and cognitive performance in an attention-based task. In two laboratory experiments, participants worked on the Stroop task, in which performance relies predominantly on attentional functions, while being exposed to different road traffic noise scenarios. These were systematically varied in macro-temporal pattern regarding break duration and distribution (regular, irregular), and played back with moderate L_{Aeq} of 42–45 dB(A). Noise annoyance ratings were collected after each scenario. Annoyance was found to vary with the macro-temporal pattern: It decreased with increasing total duration of quiet periods. Further, shorter but more regular breaks were somewhat less annoying than longer but irregular breaks. Since Stroop task performance did not systematically vary with different noise scenarios, differences in annoyance are not moderated by experiencing worsened performance but can be attributed to differences in the macro-temporal pattern of road traffic noise.

Keywords: road traffic noise; macro-temporal pattern; noise indicator; noise annoyance; cognitive performance; Stroop task; listening experiment

1. Introduction

Noise annoyance is one of the most important negative health-related effects of environmental noise [1,2]. For annoyance, exposure-response relationships are typically based on time-averaged metrics, such as the A-weighted equivalent continuous sound pressure level (L_{Aeq}), the day-night level (L_{dn}), or the day-evening-night level (L_{den}) [3–5]. However, while such noise metrics have proven to be strong predictors of annoyance (e.g., [4,6]), they ignore other potentially important acoustical and non-acoustical characteristics of a noise situation, in particular the macro-temporal pattern (e.g., [7–11]). The objective of our study therefore was to elucidate the link between the macro-temporal pattern of road traffic noise and annoyance on the one hand, and cognitive performance on the other hand, especially as the latter might moderate annoyance ratings, and because evidence of noise effects on cognitive performance is still scarce [12]. Note that the term "road traffic noise" is used throughout this paper to refer to either road traffic induced "noise" or "sound". The term "road traffic noise" is very common (e.g., [6]). However, strictly speaking, sound and noise are not the same. Sound refers to the physical quantity sound pressure from which acoustical metrics can be derived with calculations or measurements, while noise refers to unwanted sound entailing negative effects on humans (e.g., [6]). As a consequence, studies on negative effects rather refer to noise, while soundscape studies focusing on potentially positive effects refer to sound (e.g., [13]).

Road traffic noise and its effects on annoyance and cognitive performance becomes increasingly important as urbanization is progressing. While less than 34% of the global population lived in urban regions in 1960, this number rose to more than 56% globally in 2020 (and to ~74% in Europe) [14]. This growth of urban areas goes hand in hand with an increase in noise pollution, in particular due to road traffic. Accordingly, some 113 million Europeans were estimated to be exposed to road traffic noise L_{den} of 55 dB or more in 2017 [15], of which more than 72% lived in urban areas. Increasing road traffic noise calls for effective countermeasures (noise control and mitigation) to be considered by urban planners. They need to know which acoustic qualities and quantities they have to preserve or (re-)create in remnant or newly designed urban spaces. This, however, requires sufficiently funded knowledge on the effects of traffic noise. While much research was dedicated to noise annoyance in the past (e.g., [2]), effects on cognitive performance are less explored [16,17]. A recent systematic review of non-experimental studies on the association between transportation noise and cognitive performance found only 34 papers, which did not allow for a quantitative meta-analysis and were exclusively dedicated to child populations [12]. Thus, studies on mutual effects of road traffic noise on annoyance and cognitive performance of adults are desirable.

The macro-temporal pattern of noise and its effect on **noise annoyance** may be described with different indicators. The number of dominant events, typically defined relative to a threshold (e.g., Number above Threshold, NAT [18]), has been reported to be a promising predictor of annoyance [9,19,20], and also the maximum sound pressure level ($L_{\text{A,max}}$) is occasionally used for the same purpose [21]. Besides, one may use statistical levels, namely, L_{10}, L_{50} and L_{90}, to describe rare events, average noise levels and background noise [22,23], respectively, or differences between statistical levels to define fluctuation and/or emergence [24]. Further, quietness was suggested as an additional predictor for (reduced) noise annoyance [7,10]. Finally, the eventfulness of noise situations, expressed as intermittency ratio [11], was proposed as an additional indicator for annoyance. Literature indeed suggests annoyance to be associated with such indicators for the macro-temporal pattern of noise. One study found reduced annoyance in highly intermittent road traffic noise situations with only a small number of vehicles per hour [5], which might be the consequence of phases of relative quietness between events, lasting two or more minutes on average. Several other studies emphasized the need to consider quiet periods (i.e., noise breaks) in the assessment of noise impact on public health [8,25–29]. They suggested that not only the total length of noise breaks, but also their distribution and individual duration could be important [8,25,27], as longer breaks (in total and individually) might mitigate annoyance [8–10,25,27]. Here, a minimum duration of noise breaks seemed necessary to be noticeable and effective [25,27–29], which should last one minute, called "a while" ("eine Weile" in German) [25], or three minutes [27–29]. Calm periods were also found in [30] to reduce annoyance, while their pattern (regular or irregular) did not have a significant effect. However, with 0.25–1.65 s, the noise breaks were quite short. Thus, the macro-temporal pattern may be decisive for annoyance, but literature on this aspect is still quite scarce.

In addition to annoyance, the macro-temporal pattern of road traffic noise may also affect **cognitive performance**. In everyday life and at work, cognitively demanding tasks often have to be achieved in the presence of background noise. Consequently, the detrimental effects of task-irrelevant sound on cognitive performances have been explored in a multitude of basic cognitive psychological studies (see, e.g., [31–33]). However, whereas quite some research focused on chronic effects of road traffic noise on children's cognitive performance [12], surprisingly little evidence is available on acute effects on cognitive performance of adults (e.g., [17,34–37]). With regard to the macro-temporal pattern of road traffic noise as constituted by the duration and distribution of noise breaks, the effect on attentional functions is of particular interest. This is because unexpected, salient changes in the acoustic background cause the distraction of the attentional focus from the task to the background sound, so that controlled task-related processes are interrupted. This attentional capture and resulting drop in cognitive performance is known as the "deviance effect" [38]. It occurs because our auditory-cognitive system constantly monitors the acoustic background, at least to a certain extent, even when we are concentrating on a given visual cognitive task unrelated to the noise. In fact, a certain distractibility is an important prerequisite for human survival in potentially threatening environments. However, when focusing on a cognitive task, road traffic noise is arguably irrelevant in all respects. Nonetheless, its macro-temporal pattern may cause attentional capture, in particular the transitions from noisy to quiet periods and back, and/or irregular noise breaks as unanticipated changes in the auditory background. Yet while the length and distribution of noise breaks appear to affect noise annoyance, their effects on attentional capture have not been studied to our knowledge. Since subjective annoyance ratings and cognitive task performance do not necessarily go hand in hand, it is not possible to infer from noise effects on annoyance to cognitive performance effects [39–41]. Thus, both effect dimensions should be studied for a comprehensive evaluation of road traffic noise and its macro-temporal pattern, even more so as impacts on cognitive performance might moderate noise annoyance, and as mutual effects of road traffic noise have hardly been studied so far. For example, one might notice that his/her own performance is reduced under road traffic noise, and this is then expressed in a higher subjective annoyance rating.

The objective of the present study therefore was to investigate the effects of the macro-temporal pattern of different road traffic scenarios on noise annoyance and objective performance indicators of attentional functions by means of psychoacoustic laboratory experiments.

2. Methodological Approach

In this study, two experiments were conducted to investigate the effects of the two independent macro-temporal pattern variables "relative quiet time" and "quiet time distribution" (cf. Section 2.3) on short-term noise annoyance and cognitive performance in a task which predominantly relies on attentional functions: the Stroop task [42]. Experiment 1 investigated the individual and combined effects of the two variables, while experiment 2 focused on the effect of quiet time distribution in more detail. Two different versions of the Stroop task, derived from the colour test [42] and shape test [43], were used (Section 2.2). The latter were identified as suitable in a pilot study to this paper [44], where (i) the difficulty of Stroop tasks necessary for the framework of our study was assessed, (ii) interchangeable Stroop tasks were identified, and (iii) the chosen tasks were applied in a preliminary listening experiment to test their feasibility. The pilot study is described in detail in [44]. Figure 1 gives an overview of the workflow of the experiments.

In the following, Section 2.1 introduces the experimental concept of our study, Section 2.2 presents the Stroop tasks, and Section 2.3 the indicators used to quantify the macro-temporal pattern of the road traffic noise scenarios. Section 3 then documents experiment 1 and Section 4 experiment 2. Section 5 discusses the results, before Section 6 gives the major conclusions to our study.

2.1. Experimental Concept: Unfocussed Listening Experiments

In two experiments, subjectively perceived acute noise annoyance reactions (so called "short-term annoyance" [45,46] or "psychoacoustic annoyance" [47]) to road traffic noise scenarios with different macro-temporal pattern were investigated under laboratory conditions. Each scenario was several minutes long (4.5 min in experiment 1 and 10 min in experiment 2) and comprised a number of single car pass-by events.

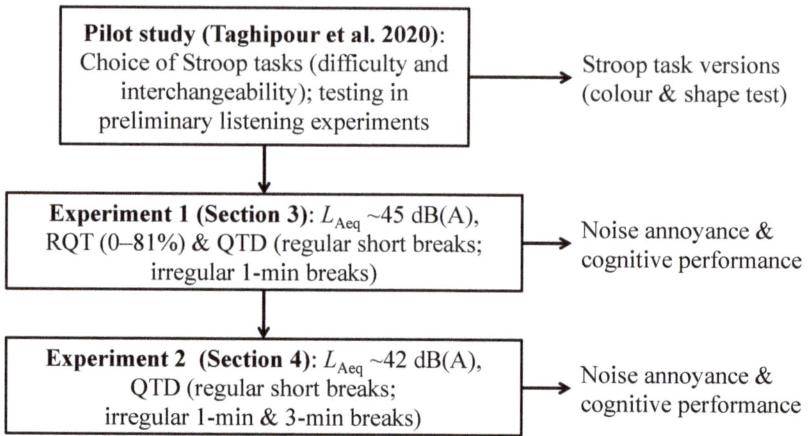

Figure 1. Study design: Pilot study to this paper by Taghipour et al. [44] to identify suitable Stroop task versions, experiment 1 on the association of noise annoyance and cognitive performance with relative quiet time (RQT) and quiet time distribution (QTD), and experiment 2 on the association with QTD. Details are given in [44] (pilot study) as well as in Sections 3 and 4 (experiments 1 and 2).

The listening experiments were designed as "unfocused listening experiments" (e.g., [48,49]), where the participants' primary focus was not on the noise scenarios but on a cognitive task (see below). While focused listening experiments are widely used in studies where participants attentively listen to and rate acoustic stimuli of relatively short duration (usually <1 min; e.g., [45,48]), unfocused experiments are typically performed for subjective assessment of noise scenarios with considerably longer durations as used here (several minutes or hours; e.g., [17,49,50]). Furthermore, the latter experimental set-ups allow both measuring the effects of sound on cognitive performance and to collect subjective annoyance (or other) ratings of the sound situations.

In the present study, the participants conducted a visually presented cognitive task, while road traffic noise scenarios were played back. The participants' primary focus was thus on the cognitive task and not on the noise scenarios. However, at the end of each noise scenario, the participants rated their noise annoyance. As laboratory setup, an office environment was chosen where an open window was simulated from which the road traffic noise would enter the office (Figure 2). To that aim, a loudspeaker playing back road traffic noise scenarios was placed in front of the closed window. For the experiments, moderate exposure scenarios with L_{Aeq} of 42–45 dB(A) were chosen, which are representative values for an office environment. The daytime limit value (impact threshold) for road traffic noise of 60 dB outdoors in residential zones according to Swiss legislation [51] and a sound level attenuation during transmission from the outside to the inside of some −15 dB for tilted windows [52,53] approximately result in the above indoor L_{Aeq}. Likewise, a road traffic noise L_{den} of 53 dB according to the recommendation of WHO [6], corresponding to a daytime L_{Aeq} of ~51 dB(A) [54], and a sound level attenuation during transmission from the outside to the inside of some −10 dB for open windows [53] lead to similar values. Besides the actual noise scenarios, constant low background sound was played back with an additional loudspeaker (cf. Section 3.1).

The experiments were approved by the ethics committee of Empa (approval CMI 2019-224 of 30 October 2019). They followed general guidelines such as [55,56] and were conducted similarly to previous experiments by the authors (e.g., [21,45]).

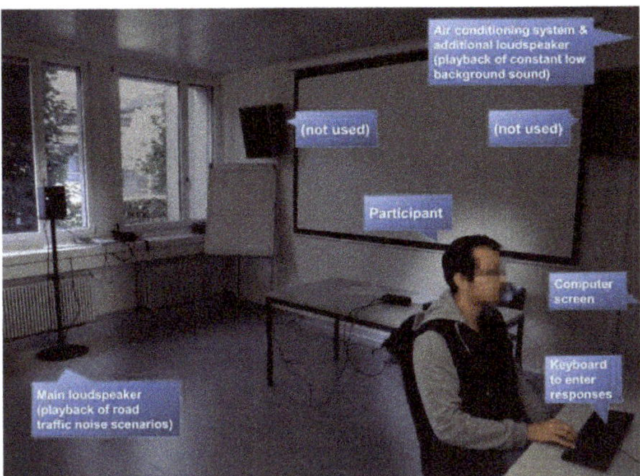

Figure 2. Photography of the laboratory setup (office environment at the Empa) used for the listening experiments. The loudspeaker positioned at the (closed) window mocked the road traffic noise scenarios at an open window, while the participant at a desk performed the visually presented cognitive task. The loudspeakers on the wall (left and right of the picture and indicated with "not used") were not used in the current experiments. For details on the air conditioning system and the additional loudspeaker see Section 3.1.

2.2. Stroop Task Versions for Unfocussed Listening Experiments

Cognitive performance was tested using different versions of the Stroop task. Details on the Stroop task are given, e.g., in [57]. In its standard version, different colour words are displayed (blue, green, red, yellow) which are either printed in the same colour as their semantic meaning (congruent item; e.g., the word "green" displayed in green colour) or in another colour (incongruent item; e.g., the word "green" displayed in blue) [42] (cf. first row of Figure 3). Participants are asked to respond to the colour in which the word is printed (in the latter example: blue) and not the word's semantic (here: green). Reading the semantics of a word is an automated process for skilled readers, so that in the case of incongruent items the automatically activated word must be inhibited and the correct response–namely, the print colour of the word–must be specifically selected. Therefore, an increase in errors and/or response times occurs for incongruent items compared to congruent items, which is the so-called Stroop effect [42,58].

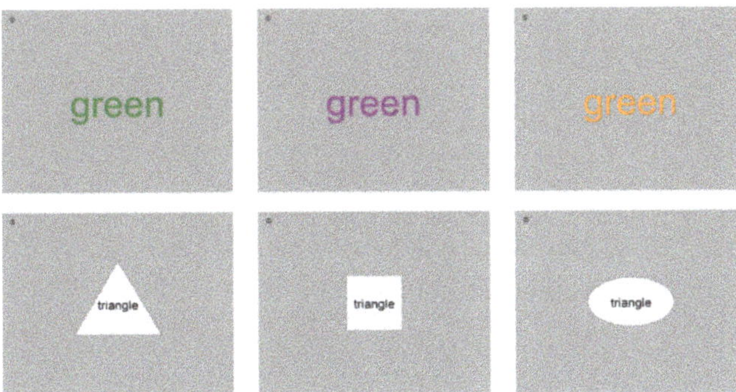

Figure 3. Screenshot examples of the colour version (**top**) and the shape version (**bottom**) of the Stroop task used in this study. Congruent items (the two stimulus' attributes match) are shown left and incongruent items (stimulus' attributes do not match) in the middle and right.

Performance in the Stroop task relies on attentional functions, namely, selective attention and inhibitory functions, so that it should be sensitive to attentional capture induced by transitions from a quiet period to road traffic noise or vice versa. As working on a large amount of look-alike items for prolonged time periods might become too tiresome, different versions of the Stroop task were used in the present study. Two versions of the Stroop task were identified in a pilot experiment to this study (details see [44]) as sufficiently equivalent with respect to difficulty, interchangeability and observability of the aforesaid Stroop effect (cf. Figure A1 in the Appendix A). The first version was a colour test where, contrary to its standard version ([42], see above), participants were asked for the semantics of the colour word (instead of its actual print colour) (cf. first row of Figure 3). The second version was a shape test (cf. [43]), where participants were asked to identify the shape of a geometric form, while a written word within specified the same or a different geometric form (cf. second row of Figure 3). Here, congruent items are those in which the semantic meaning of the word and geometric shape match (e.g., the word "rectangle" is printed in a rectangle), while these do not match for incongruent items (e.g., the word "rectangle" is printed in a circle while the latter should be named).

In addition to the above two versions of the Stroop Tasks, two variants each were used to keep the task to be processed sufficiently diverse:

- Shape test variant A: oval, square, and triangle (cf. Figure 3)
- Shape test variant B: circle, rectangular, and star
- Colour test variant A: green, orange, and purple (cf. Figure 3)
- Colour test variant B: blue, yellow, and red

The different versions/variants of the Stroop task were implemented in a listening test program in the Python-based PsychoPy software environment [59]. The individual trials were presented on a monitor screen, and responses were given by the participants on a keyboard and stored by the program.

2.3. Indicators for the Macro-Temporal Pattern of the Road Traffic Noise Scenarios

The following indicators were used to quantify the macro-temporal pattern of the road traffic noise situations: Number of events (N), Relative Quiet Time (RQT), Intermittency Ratio (IR), Centre of Mass Time (CMT), A-weighted and FAST-time weighted maximum sound pressure level (L_{AFmax}), and Quiet Time Distribution (QTD). The indicators were all calculated from the road traffic noise scenarios (see below) in MATLAB Version 2019a (The MathWorks, Inc., Natick, MA, USA).

Number of events (N): Since this study used isolated car pass-by events mixed to prepare the scenarios (see below), the number of events, as well as the logarithm $\log(N)$ as sometimes used to predict annoyance (e.g., [20]), in each scenario were directly available.

Relative Quiet Time (RQT): Based on suggestions by [10], RQT is determined as the ratio of total duration of quiet periods (T_{quiet}) to total duration of a scenario ($T_{scenario}$) [26]. To that aim, T_{quiet} is calculated as the sum of all (individual) quiet periods and divided by $T_{scenario}$ as

$$\text{RQT (\%)} = 100 \cdot \frac{T_{quiet}}{T_{scenario}} \tag{1}$$

Intermittency Ratio (IR, %): IR is a measure for the eventfulness of a noise scenario [11]. It expresses the proportion of the acoustical energy of all individual noise events relative to the total sound energy of a scenario as

$$IR \text{ (\%)} = 100 \cdot 10^{0.1(L_{Aeq,T,\text{Events}} - L_{Aeq})}, \tag{2}$$

where $L_{Aeq,T,\text{Events}}$ is calculated from contributions of events exceeding a given threshold K. In contrast to other descriptors working with thresholds, the latter is not constant, but defined dynamically relative to the L_{Aeq} of the scenarios using

$$K = L_{Aeq} + C[\text{dB}], \tag{3}$$

where C is a constant offset, set to 3 dB. *IR* ranges from 0–100%. An *IR* larger than 50% indicates that more than half of the total sound energy is due to distinct pass-by events. In situations where all events clearly emerge from background noise (e.g., at a receiver close to a railway track), *IR* gets close to 100%, while constant road traffic as observed from a receiver not too close to a motorway yields only small *IR* values. Note that while a high *IR* is a precondition for noise breaks (large RQT) to occur, it does not allow studying the effect of QTD (i.e., the temporal distribution and length of the noise breaks).

Centre of Mass Time (CMT): CMT is an indicator for quiet periods which penalizes the fragmentation of quiet periods and rewards their clustering and thus increases with longer quiet time periods [8]. It is calculated as

$$\text{CMT}\ (s) = \frac{\sum t_i{}^2}{\sum t_i}, \tag{4}$$

where t_i is the duration of the i-th (individual) quiet period in the scenario (in seconds).

Quiet Time Distribution (QTD): QTD is a categorical variable for the nature of noise breaks. Here, it discriminates between regular and irregular temporal distribution of the breaks as well as between different durations of the irregular noise breaks.

3. Experiment 1

In experiment 1, the individual and combined effects of the independent macro-temporal pattern indicators RQT and QTD on noise annoyance and cognitive performance in the Stroop task were investigated.

3.1. Audio Processing and Resulting Road Traffic Noise Scenarios

Road traffic noise scenarios (WAVE PCM format) were prepared in MATLAB Version 2019a (The MathWorks, Inc., Natick, MA, USA) from stereo recordings with a Jecklin disk setup made within a previous study [45], of individual car pass-by events which were dominated by tire/road noise. Since the laboratory setup should represent an office environment in which the road traffic noise enters through an open window, the signals were down-mixed from stereo to mono by means of crossfading. The recordings, processing, and playback was carried out at a sampling frequency of 44.1 kHz.

Road traffic noise scenarios were created from excerpts of the individual car pass-by events by mixing them together sequentially (and sometimes slightly overlapping) in time. After careful inspection of the audio files (audibly as well as based on their A-weighted and FAST-time weighted level-time histories, L_{AF}), an average duration of 10 s was chosen for the excerpts. However, to obtain realistic sound scenarios, three excerpts, of 9, 10, and 11 s length, were cut from each signal. One of these three excerpts per event was randomly chosen for the preparation of a scenario. The excerpts were gated with raised-cosine ramps of 2 s. They were further highpass and lowpass filtered at 52 Hz and 10 kHz, respectively, to consider the limits of the loudspeaker at low frequencies and inherent recording noise at high frequencies. In total, seven scenarios, each lasting 4.5 min, were prepared for experiment 1. Additionally, two 30 s long road traffic noise scenarios were created for the participant's familiarization period with the noise and the cognitive task at the beginning of the experimental session.

The road traffic noise scenarios covered four levels of RQT, namely, 0.0% (corresponding to 36 car pass-by events), 44.3% (15 events), 62.9% (10 events), and 81.5% (5 events). Further, two types of QTD were used for the quiet periods: either a regular distribution (referred to as "regular" in the following account) or a combination of short quiet periods and two longer (1-min) quiet periods (referred to as "irregular"). While the situation with 0.0% RQT served as a reference without quiet periods, the three levels of RQT (44.3%, 62.9%, 81.5%) were combined with the two QTD types, (total of $3 \times 2 + 1 = 7$ road traffic noise scenarios). All road traffic noise scenarios had the same L_{Aeq} of 54 dB(A) at the window (measured 50 cm away from and in front of the loudspeaker) and of 44.5 dB(A) at the participant's ear level at the desk. As the number of car pass-by events varied between scenarios, the L_{Aeq} of the individual pass-by events had to be adjusted. Figure 4 shows the level-time histories of the road traffic noise scenarios, visualizing different distributions and resulting lengths of the quiet periods, and Figure 5 the corresponding one-third octave spectra, which were all very similar. Table 1 presents the indicators for the resulting macro-temporal pattern of the scenarios, and Table A1 in the Appendix A presents the correlation analysis using Spearman's rank correlation coefficient (r_s) [60] for the continuous indicators, as a measure of similarity of the indicators without an a priori assumption of a linear relation. While the $L_{AF,max}$ generally decreases with increasing number of events to obtain the same overall L_{Aeq} for all scenarios, a few events of scenarios S5 and S6 (each encompassing 15 events) had a similar $L_{AF,max}$ as the events of S3 and S4 (each encompassing 10 events), so that the $L_{AF,max}$ were almost identical for those four scenarios (Table 1). N, RQT, *IR* and $L_{AF,max}$ were closely correlated to each other. CMT, in contrast was not correlated to these indicators (Table A1), but was closely related to QTD, with substantially larger values for irregular than for regular distributions (Table 1). Thus, with N, *IR* and $L_{AF,max}$ being closely related to RQT and CMT being closely related to QTD, the association of the macro-temporal pattern with annoyance and cognitive performance was mainly investigated with RQT and QTD (cf. Sections 3.4 and 3.5).

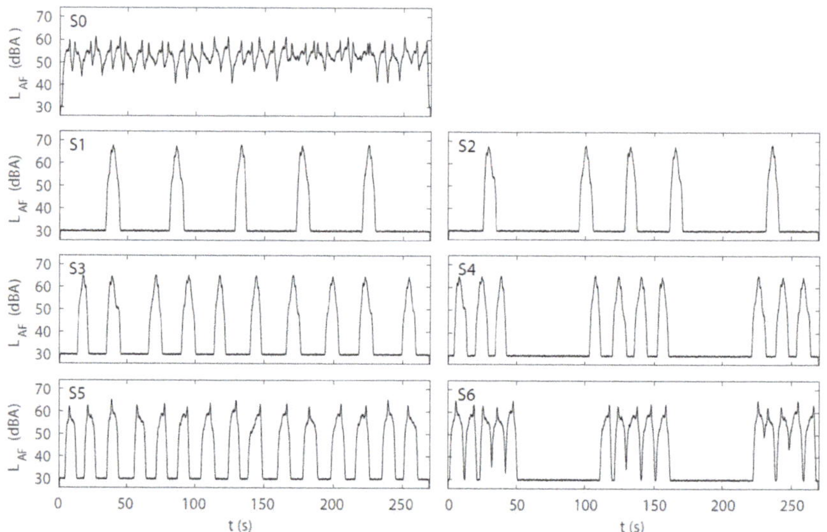

Figure 4. A-weighted and FAST-time weighted level-time histories (L_{AF}) of the road traffic noise scenarios in experiment 1. S0–S6 refer to scenario 0 (reference) to 6 (cf. Table 1).

Note that in addition to these road traffic noise scenarios, the participants were exposed to a constant background sound with an L_{Aeq} of 30 dB(A), which was a combination of filtered pink noise (played back via an additional loudspeaker) and sound from a low-level running office air conditioning system. The additional loudspeaker was located at the wall in front of and above the participant, at the same height as the running low-level office air-conditioning system, so that both sounds were received from roughly same direction and combined to one background sound source. The background sound helped masking possible low-level sounds from outside the office environment, which was not an isolated listening booth. In addition, a sign was put up during the experiments in the corridor outside the office, asking passers-by to be silent. Thus, sounds from outside the office were minimized. With the played-back background sound being constant and ~15 dB lower than the actual road traffic noise scenarios, both sound sources (sound outside the office and background sound) are negligible as a source of bias for the annoyance ratings. Also, even if the background sound within the mock office would have somewhat affected the participants' perception and/or performance, this is something that would also be present in a real office environment.

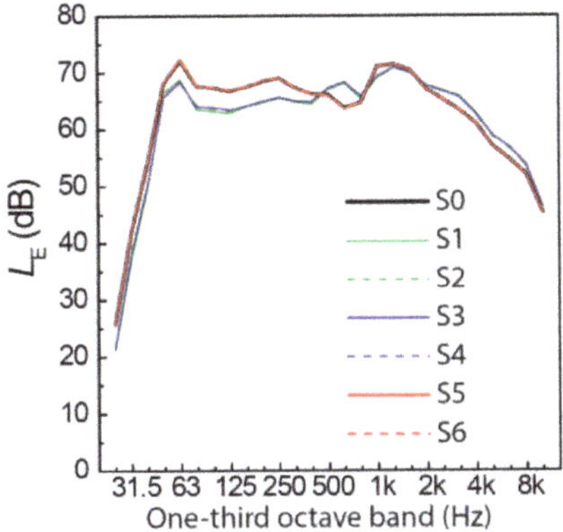

Figure 5. One-third octave spectra of the road traffic noise scenarios in experiment 1. S0–S6 refer to scenario 0 (reference) to 6 (cf. Table 1).

Table 1. Characterization of the macro-temporal pattern of the road traffic noise scenarios of experiment 1 (N = number of events, RQT = Relative Quiet Time, QTD = Quiet Time Distribution, IR = Intermittency Ratio, CMT = Centre of Mass Time, $L_{AF,max}$ = maximum sound pressure level as maximum of the whole traffic noise scenario).

Scenario No.	N	RQT (%)	QTD	IR (%)	CMT	$L_{AF,max}$ (dB(A))
S0	36	0.00	not applicable	30.4	Not applicable	61.4
S1	5	81.4	regular (shorter breaks)	94.4	36.7	67.9
S2	5	81.4	irregular (incl. 2 × 1-min breaks)	94.8	44.3	68.2
S3	10	62.9	regular shorter breaks)	89.5	15.5	65.2
S4	10	62.9	irregular (incl. 2 × 1-min breaks)	90.0	43.9	65.2
S5	15	44.3	regular (shorter breaks)	74.3	7.7	65.2
S6	15	44.3	irregular (incl. 2 × 1-min breaks)	76.0	60.3	65.2

3.2. Experimental Procedure

The experiments were conducted in single sessions in English. To ensure sufficient understanding of the experimental tasks, one requirement for study participation was to have good self-reported English language skills. In addition, after task instruction the participants could ask the experimenter in case of ambiguities.

Participants first answered questions about their hearing status, vision, and well-being for inclusion and exclusion criteria, which were (i) self-reported normal hearing (not hearing impaired), (ii) self-reported normal or corrected-to-normal vision (but not colour blind), (iii) legal age (18 years or older) and (iv) feeling well (not further specified). Thereafter, they read instructions on the road traffic noise scenarios, the cognitive task and the test program. To familiarize them with the two versions of the Stroop task, the two short road traffic noise scenarios were used: Participants worked on trials of the colour version of the Stroop task during the first short scenario and of the shape version during the second one. Then, data collection in the actual listening experiment started. During each noise scenario, the participant worked on trials of one version of the Stroop task for the first 135 s and then of the other version for the second 135 s. Congruent and incongruent trials were presented in random order. An overall mixing ratio of approximately 50% each was secured by the program increasing the probability of drawing either the congruent or incongruent trials after 60% of a noise scenario's duration. Participants were asked to respond to the semantics of the colour word (colour version) or the shape of the geometric form (shape version) as fast and as accurately as possible. Immediately after the participant's response (without any time delay), the next trial started automatically. There was only a break in Stroop tasks between the noise scenarios, when no sound was played back. The participants did the Stroop task self-paced, which resulted in a different number of trials per participant and noise scenario, depending on how fast they worked on the tasks.

The sequence of the two Stroop versions was randomized for each noise scenario, as was the sequence of the noise scenarios. After each noise scenario, participants answered the following question, which was adapted from the ICBEN noise annoyance question [3,61]: "What number from 0 to 10 represents best how much you were bothered, disturbed, or annoyed by the sound?" The participants gave their rating by means of a slider in the test program on the unipolar numerical ICBEN 11-point scale. As the spacing of the 11-point scale is equal (and thus interval-scaled), it allows treating the data as continuous in statistical analyses, even though by definition the scale is ordinal [3]. This is supported by literature, given that the ordinal variable has five or more categories [62–64].

After a break of 30 s the next noise scenario started. The total experiment lasted approximately 50 min, with the actual unfocussed listening test taking around 35 min.

3.3. Participants

The participants were mostly recruited within Empa, via internal online advertisement or direct verbal recruitment. Twenty-four persons (11 females and 13 males), aged between 19 and 63 years (median of 28.5 years), participated in experiment 1. This number of participants lies well within the range of 16–32 participants proposed in [55] to obtain reliable experimental results. All participants fulfilled the requirements for participation (self-reported normal hearing, self-reported normal or corrected-to-normal vision, not colour blind, legal age and feeling well, see above). Written consent for participation was collected from all participants.

3.4. Data Analysis

Annoyance: In total 168 annoyance ratings were obtained (i.e., 24 participants $\times$ 7 road traffic noise scenarios).

Performance: Task completion was self-paced, i.e., each participant had an individual pace in completing the tasks. This resulted in different amounts of worked-out trials per noise scenario and participant. On average, 208 trials in the Stroop task were worked-out, ranging from 85–265 trials per participant and traffic noise scenario, meaning that the slowest participant completed 82 trials during one specific noise scenario, and the fastest participant 262 trials during one specific noise scenario. In sum, a total of 34,911 individual responses (trials) were available and processed as follows.

Reaction times (RTs; in ms): Each trial not correctly worked-out counted as an error. As usual in analysis of RTs, error trials were removed from the data set, as cognitive mechanisms might have been different from those involved in successful task processing. In a second step, long RTs (exceeding 2 standard deviations of mean overall RTs of the experiment, corresponding to RTs > 1771 ms) were removed, as again other mechanisms might have played a role (e.g., the participant re-reading the instructions on the task or accidentally pressing a response key). In total, 3000 individual responses (trials) (9.1%) were removed. In a last step, the remaining 31,911 individual responses were averaged per participant and road traffic noise scenario separately for congruent and for incongruent trials to obtain mean RTs (data set with a total of 336 entries).

Error rate (ER; in %): In a first step, individual colour and shape task versions/variants (cf. Section 2.2) per participant with too high rates of wrong answers (namely, ER > 10%) were removed, as these tasks were likely misunderstood by the participants (e.g., answering the colour instead of the required semantics of the word). In total, 3,410 trials (9.8%) were thus removed. The remaining 31,501 individual trials were again averaged per participant and noise scenario separately for congruent and incongruent trials to obtain the mean ERs (data set with a total of 336 entries).

The data was statistically analysed, separately for annoyance on the one hand, and RT and ER as measures of cognitive performance on the other hand. To that aim, linear mixed-effects models were established (see, e.g., [65]). These models allow separating fixed effects (here, the variables RQT and QTD, which were correlated with the other indicators, cf. Section 3.1) and random effects (the participants, modelled with a simple random intercept: one for each participant). Further, the playback number (i.e., the serial position with which the noise scenarios had been played) was included to test for order effects [66]. The statistical analysis was done with IBM SPSS Version 25 using the procedure MIXED.

3.5. Results

3.5.1. Annoyance

Table 2 shows the correlations (Spearman's rank correlation coefficient r_s [60] and Pearson's r, the latter assuming a linear relation) of the annoyance ratings with the continuous indicators for the temporal pattern. Both correlation analyses reveal the same insights, although correlation with Spearman's r_s is less strong than with Pearson's r. Annoyance increased with increasing N (more events) and CMT (i.e., longer noise breaks, indicating irregular distribution of the events), but decreased with increasing RQT (longer total quiet time), IR (increasingly dominant, here meaning less, single events) and $L_{AF,max}$ (louder, here meaning less, events). As the acoustical indicators are closely correlated to either CMT or QTD (cf. Table A1), the following account focusses on RQT and QTD. As Table 2 reveals, the correlations are rather moderate. One reason for this is that the correlation analysis was performed for the individual annoyance data (168 ratings: cf. Section 3.4) without accounting for individual differences between participants' ratings. This shortcoming is overcome by the subsequent hierarchical mixed-effects models, where the participants are modelled with a random intercept.

Table 2. Correlation analysis: Spearman's rank correlation coefficient r_s [60] and Pearson's r for correlations between annoyance and the indicators for the macro-temporal pattern of the road traffic noise scenarios (cf. Section 2.3) (N = number of events, RQT = Relative Quiet Time, IR = Intermittency Ratio, CMT = Centre of Mass Time, $L_{AF,max}$ = maximum sound pressure level).

Correlation	N	Log(N)	RQT (%)	IR (%)	CMT	$L_{AF,max}$ (dB(A))
Spearman's r_s	0.14 †	0.14 †	−0.14 †	−0.10	0.15 †	−0.15 *
Pearson's r	0.22 **	0.18 **	−0.20 **	−0.23 **	0.15 *	−0.16 **

† $p < 0.08$, * $p < 0.05$, ** $p < 0.01$.

Figure 6 shows the association of annoyance with RQT and QTD. RQT increasing from 0% to 44–81% was associated with decreased annoyance. QTD was linked with annoyance as well, with regular breaks being less annoying than irregular breaks. An interaction between RQT and QTD was not observable (Figure 6c). Besides, annoyance increased with playback number increasing from 1–7 (not shown). This simple order effect was expected and observed in other studies by the same authors (e.g., [21,45]), indicating that the participants got increasingly annoyed by the road traffic noise scenarios over time.

Linear mixed-effects modelling analysis confirmed these observations and significant differences between regular and irregular QTD (cf. Figure 6b,c). Here, two models are reported, which either relate annoyance to RQT (model M_{RQT}) or to QTD (model M_{QDT}). The first model, M_{RQT}, reveals the dependence of annoyance on the continuous variables RQT and playback number (PN). This model takes into account all noise scenarios, S0–S6.

$$Annoy_k = \mu + \beta_1 \times RQT + \beta_2 \times PN_k + u_k + \varepsilon_k. \tag{5}$$

In Equation (5), *Annoy* is the dependent variable annoyance, μ denotes the overall grand mean, β_1 and β_2 are regression coefficients for the continuous variables RQT and PN, respectively, of the seven scenarios (S0–S6), u is the participants' random intercept ($k = 1$–24), and the error term ε is the random deviation between observed and expected values of *Annoy*. Table 3 gives the model coefficients. The model M_{RQT} shows that annoyance significantly decreases by 1.4 units on the 11-point scale when RQT increases from 0–81% (cf. Figure 6a), and significantly increases by 1.4 units with a playback number increase from 1–7 (incidentally a very similar increase as for RQT increasing from 0–81%).

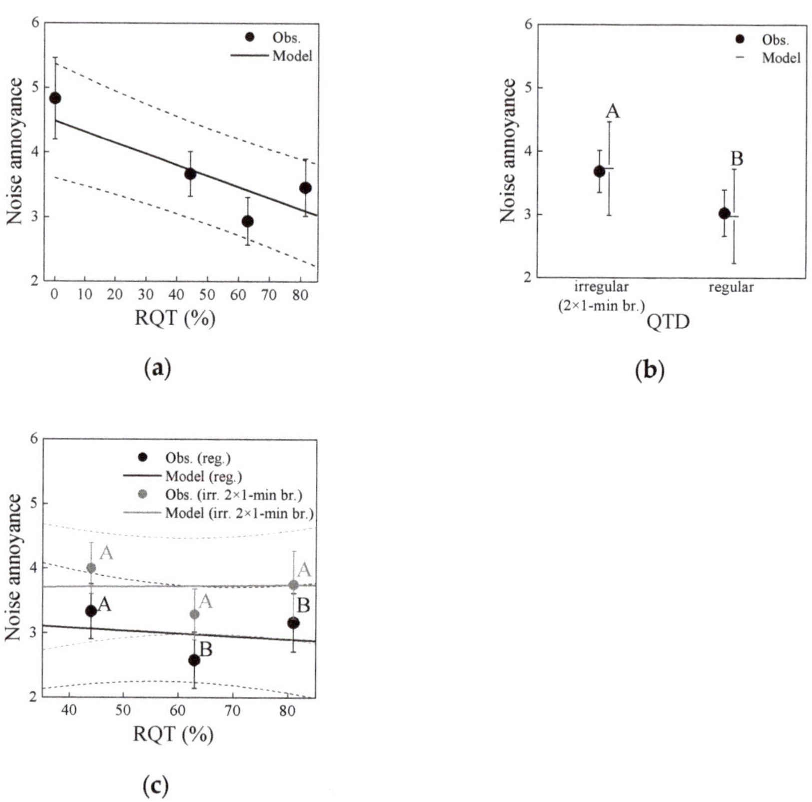

Figure 6. Noise annoyance as a function of (**a**) relative quiet time (RQT), (**b**) quiet time distribution (QTD) and (**c**) both RQT and QTD as found in experiment 1. Circles represent mean observed values (Obs.) with standard error bars, and lines the corresponding mixed-effects models with 95% confidence intervals, in (**b**) as horizontal lines with confidence intervals. In (**b**,**c**), significant differences between estimated marginal means ($p < 0.05$; pairwise comparisons with Bonferroni correction) of regular and irregular QTD are indicated by differing letters.

Table 3. Model coefficients (Coeff.), 95% confidence intervals (CI) and probability values (p) of the linear mixed-effects model M_{RQT} for annoyance (parameters and symbols: Equation (5)).

Parameter	Symbol	Coeff.	95% CI of Coeff.	p
Intercept	μ	3.581	[2.558; 4.604]	<0.001
RQT (%)	β_1	−0.017	[−0.027; −0.008]	<0.001
Playback No. (PN)	β_2	0.227	[0.105; 0.349]	<0.001
Random effect variance	$u^2{}_k$	2.559	[2.028; 3.229]	<0.001
Residual variance (intercept)	$\varepsilon^2{}_k$	2.734	[1.419; 5.268]	0.003

The second model, M_{QDT}, reveals how annoyance is linked to QTD. In this model, only six scenarios, S1–S6, are taken into account, since no level of QTD is applicable for S0 with RQT of 0% (cf. Table 1). In the absence of S0, RQT is not linked to annoyance ($p > 0.8$; also obvious in Figure 6c). Also, there was no significant interaction between RQT and QTD ($p > 0.7$; cf. Figure 6c). Model M_{QDT} therefore reduces to

$$Annoy_{ik} = \mu + \tau_{QTD,i} + \beta \times PN_{ik} + u_k + \varepsilon_{ik}. \tag{6}$$

In Equation (6), τ_{QTD} is the categorical variable QTD (2 levels: $i = 1, 2$ for regular and irregular) of the six scenarios (S1–S6), and the other variables have the same notation as in Equation (5). Table 4 gives the model coefficients. According to model M_{QTD}, annoyance is significantly higher for longer, irregular than for shorter, regular breaks, but the difference of ~0.7 points on the 11-point scale is moderate (cf. Figure 6b). Further, annoyance significantly increases with playback number (as in above model M_{RQT}).

Table 4. Model coefficients (Coeff.), 95% confidence intervals (CI) and probability values (p) of the linear mixed-effects model M_{QTD} for annoyance. (Parameters and symbols: Equation (6)).

Parameter	Symbol	Coeff.	95% CI of Coeff.	p
Intercept	μ	2.181	[1.276; 3.086]	<0.001
QTD	$\tau_{QTD,i=irreg}$ (2 × 1-min)	0.747	[0.253; 1.242]	<0.005
	$\tau_{QTD,i=reg}$	0 [a]		
Playback No. (PN)	β	0.200	[0.073; 0.327]	<0.005
Random effect variance	$u^2{}_k$	2.211	[1.713; 2.854]	<0.001
Residual variance (intercept)	$\varepsilon^2{}_{ik}$	2.417	[1.240; 4.713]	0.003

[a] Reference.

3.5.2. Cognitive Performance

Performance data was first checked for the Stroop effect with a simple model considering congruency as the sole fixed effect. In fact, the Stroop effect was found for both, RTs and ERs: Overall, the effect of congruency was highly significant for RTs ($p < 0.001$), with incongruent trials (mean RT = 682 ms; standard deviation SD = 148 ms) being answered 31 ms (or 5%) slower than congruent trials (mean RT = 652 ms, SD = 138 ms), as usual in the Stroop paradigm. Furthermore, the Stroop effect was also found for ERs ($p < 0.05$), with more errors been made in incongruent trials (mean ER = 2.4%, SD = 2.6%) than in congruent trials (mean ER = 2.0%, SD = 2.2%). Consequently, the effects of the different road traffic noise scenarios on RTs and ERs were analysed separately for congruent and incongruent trials in the following.

RT: Figure 7 shows the association of RT with RQT and QTD, separately for congruent and incongruent trials in the Stroop task. RT was not linked to RQT, except that it tended to be somewhat longer for the longest RQT (81%) than the other RQTs (0–63%) (Figure 7a). RT, however, was linked to QTD, being somewhat longer for regular than irregular breaks (Figure 7b). Congruent and incongruent stimuli were affected similarly strong. Besides, RT decreased with increasing playback number (not shown) as participants got quicker with answering the trials of the Stroop task over time, indicating that they got increasingly practiced.

Linear mixed-effects model analysis again confirmed these observations and significant differences between regular and irregular QTD (cf. Figure 7b). It revealed that RT was not significantly associated with RQT for incongruent ($p = 0.29$) and congruent trials ($p = 0.65$) (cf. Figure 7a), but with QTD (p's < 0.05; Figure 7b) and playback number (p's < 0.001) for both incongruent and congruent trials (details not shown). While the effect of QTD was significant, it was quite small (less than 30 ms compared to overall ~650 ms RTs on average, corresponding to a relative change of less than 5%; cf. Figure 7b). RT decreased by some 140 and 130 ms for incongruent and congruent trials, respectively, with playback number increasing from 1–7.

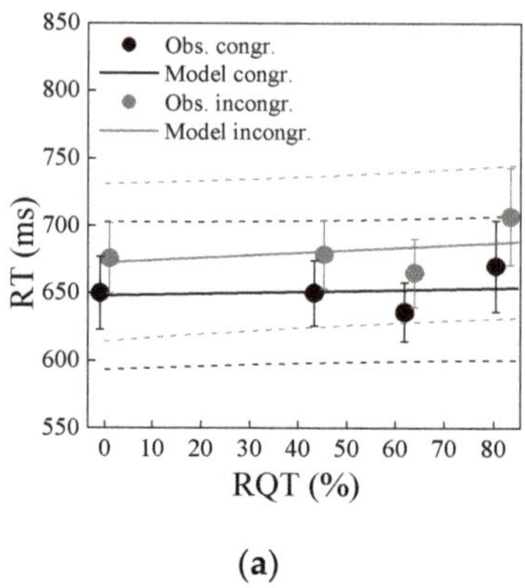
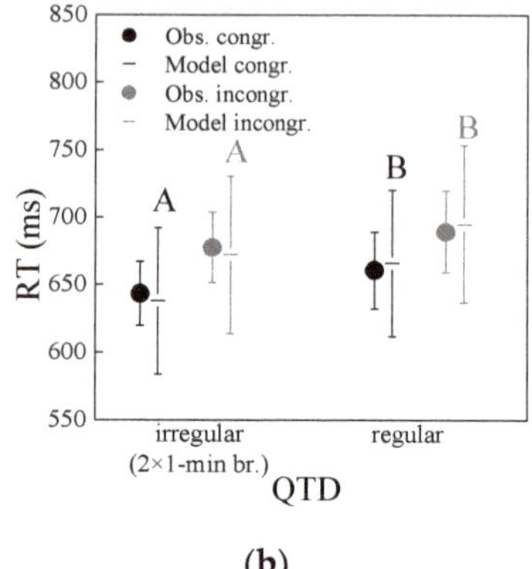

(**a**)   (**b**)

Figure 7. Reaction time (RT) as a function of (**a**) relative quiet time (RQT) and (**b**) quiet time distribution (QTD), for the congruent and incongruent items of the Stroop task in experiment 1. Circles represent mean observed values (Obs.) with standard error bars, and lines the corresponding mixed-effects models with 95% confidence intervals, in (**b**) as horizontal lines with confidence intervals. In (**b**), significant differences between estimated marginal means ($p < 0.05$; pairwise comparisons with Bonferroni correction, performed separately for the congruent and incongruent items) of regular and irregular QTD are indicated by differing letters.

ER: In both incongruent and congruent trials, ER varied neither with RQT nor with QTD nor with playback number (not shown), as also confirmed by mixed-effects model analysis (p's > 0.30 for RQT, p's > 0.26 for QTD, p's > 0.23 for playback number).

4. Experiment 2

In experiment 2, the effects of QTD were explored in more detail. A new sample of volunteers was recruited; no one participated in both experiments.

4.1. Audio Processing and Resulting Road Traffic Noise Scenarios

Three road traffic noise scenarios (WAVE PCM format) were again prepared in MAT-LAB Version 2019a (The MathWorks, Inc., Natick, MA), in the same way and from the same recordings as in experiment 1. Furthermore, participants were also exposed to the same constant background sound at an L_{Aeq} of 30 dB(A) (Section 3.1). Each of the three noise scenarios was 10 min long. For training, the same two 30 s long noise scenarios as in experiment 1 were used.

The three road traffic noise scenarios had the same RQT and $L_{AF,max}$ of the individual car pass-by events, but differed with respect to QTD. Three levels of QTD were used: regular quiet periods, a combination of short quiet periods and six 1-min quiet periods, or two 3-min quiet periods ("irregular"). Each noise scenario contained 25 car pass-by events. The scenarios had an L_{Aeq} of 51 dB(A) at the window (measured 50 cm away from and in front of the loudspeaker) and of 41.5 dB(A) at participant's ear level at the desk. Figure 8 shows the level-time histories of the scenarios with different QTDs and resulting lengths of the noise breaks, and Figure 9 their corresponding one-third octave spectra, which were all identical because the same individual car pass-by events were used to generate the three scenarios. Table 5 presents the indicators for the resulting macro-temporal pattern of the scenarios. Here, the association of the macro-temporal pattern with annoyance and cognitive performance was mainly investigated with QTD (as CMT was closely related to QTD, cf. Section 3.1), while RQT, N, and $L_{AF,max}$ were the same for S1–S3 and IR varied only little (Table 5).

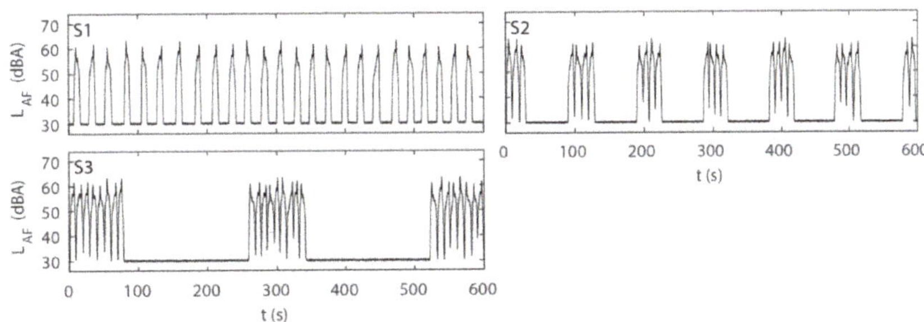

Figure 8. A-weighted and FAST-time weighted level-time histories (L_{AF}) of the road traffic noise scenarios in experiment 2. S1–S3 refer to noise scenario 1–3 (cf. Table 5).

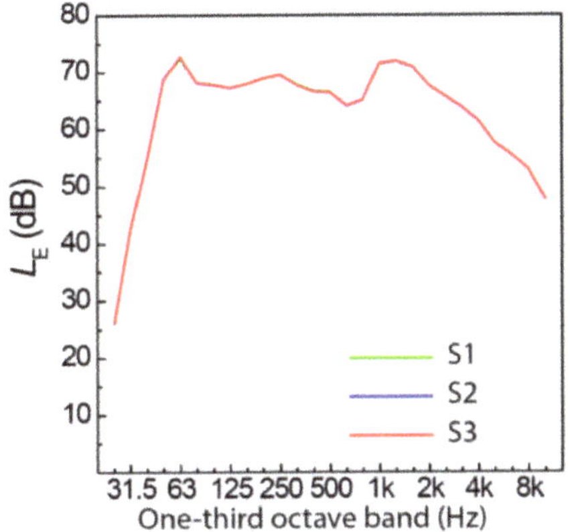

Figure 9. One-third octave spectra of the road traffic noise scenarios in experiment 2. S1–S3 refer to noise scenario 1–3 (cf. Table 5). Note that the three spectra are identical because the same car pass-by events were used to generate the three scenarios.

Table 5. Characterization of the macro-temporal pattern of the road traffic noise scenarios of experiment 2 (N = number of events, RQT = Relative Quiet Time, QTD = Quiet Time Distribution, IR = Intermittency Ratio, CMT = Centre of Mass Time, $L_{AF,max}$ = maximum sound pressure level as maximum of the whole traffic noise scenario).

Scenario No.	N	RQT (%)	QTD	IR (%)	CMT	$L_{AF,max}$ (dB(A))
S1	25	58.3	regular	82.4	13.7	63.4
S2	25	58.3	irregular (incl. 6 × 1-min breaks)	83.7	61.7	63.4
S3	25	58.3	irregular (incl. 2 × 3-min breaks)	83.2	185.1	63.4

4.2. Experimental Procedure

The procedure of experiment 2 closely followed that of experiment 1. Experiment 2 was conducted in single sessions in English. It lasted 45–50 min, with the actual unfocused listening test taking around 32 min.

4.3. Participants

The participants were again mostly recruited within Empa, via internal online advertisement or direct verbal recruitment. Twenty-five persons (12 females and 13 males), aged between 26 and 61 years (median of 33.0 years) participated in experiment 2. All participants fulfilled the requirements for participation (self-reported normal hearing, self-reported normal or corrected-to-normal vision, not colour blind, legal age and feeling well; cf. Section 3.2). Written consent was collected from all participants.

4.4. Data Analysis

Annoyance: In total, 75 annoyance ratings (25 participants $\times$ 3 traffic noise scenarios) were obtained.

Performance: Since task completion was self-paced, different amounts of worked-out trials resulted per participant and road traffic noise scenario. On average, 452 trials in the Stroop tasks were worked-out, ranging from 301–593 trials per participant and noise scenario. In total, 33,915 individual responses (trials) were available and processed analogously as in experiment 1 (Section 3.4), removing error trials as well as RTs exceeding 2 standard deviations of mean overall RTs, corresponding to RTs > 1724 ms. Thus, 2688 individual trials (8.3%) were removed for RT analysis. For ER analysis, 3153 individual trials (9.3%) of task versions/variants with too high rates of wrong answers (again, ER > 10%) were removed to ensure sufficient task understanding. The remaining 31,227 (RT) and 30,762 individual trials (ER) were then averaged per participant, noise scenario and congruency (congruent/incongruent trials) to obtain the mean RTs (in ms) and ERs (in %) (data set with a total of 150 entries).

As in experiment 1, the data was statistically analysed with linear mixed-effects models, separately for annoyance, RT and ER. As fixed effects, QTD as well as the playback number were used, and as random effects the participants (simple random intercept). The statistical analysis was again performed with IBM SPSS Version 25 using the procedure MIXED.

4.5. Results

4.5.1. Annoyance

Figure 10 shows the association of annoyance ratings with QTD. In line with experiment 1 (Figure 6b), annoyance was associated with QTD. The longest (3-min) breaks were somewhat more annoying than shorter breaks (irregular 1-min or even shorter, regular breaks). In contrast to experiment 1, however, the shorter irregular 1-min breaks were associated with very similar mean annoyance ratings as the regular breaks.

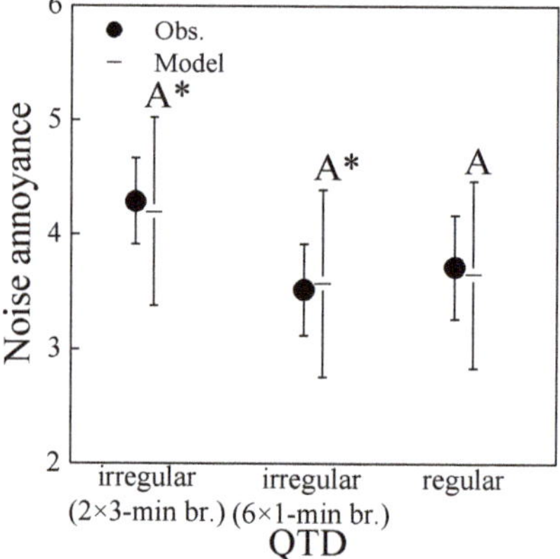

Figure 10. Noise annoyance as a function of quiet time distribution (QTD) in experiment 2. Circles represent mean observed values (Obs.) with standard error bars, and horizontal lines the corresponding mixed-effects model with 95% confidence intervals. Significant differences between estimated marginal means ($p < 0.05$; pairwise comparisons with Bonferroni correction) of different QTDs would be indicated by differing letters (* trend, $p = 0.06$ between irregular breaks).

In line with these observations, linear mixed-effects model analysis (Table 6), using the approach of Equation (6) (model M_{QDT}, but with τ_{QTD} with 3 levels, $i = 1$–3, for regular and irregular with 1-min or 3-min breaks), revealed that the overall association of annoyance with QTD was not significant ($p = 0.13$). In fact, only the annoyance to the 3-min and 1-min irregular breaks was in tendency different by ~0.6 units on the 11-point scale ($p = 0.06$; Figure 10). Again, playback number was significantly linked to annoyance ($p < 0.001$).

Table 6. Model coefficients (Coeff.), 95% confidence intervals (CI) and probability values (p) of the linear mixed-effects model M_{QDT} for annoyance in experiment 2. The parameters and symbols are explained in Equation (6) of experiment 1 (but with τ_{QTD} with 3 levels).

Parameter	Symbol	Coeff.	95% CI of Coeff.	p
Intercept	μ	2.442	[1.378; 3.505]	<0.001
	$\tau_{QTD,\ i=\text{irreg}\ (2\ \times\ 3\text{-min})}$	0.546	[−0.113; 1.205]	=0.10
QTD	$\tau_{QTD,\ i=\text{irreg}\ (6\ \times\ 1\text{-min})}$	−0.079	[−0.732; 0.573]	=0.81
	$\tau_{QTD,\ i=\text{regular}}$	0 [a]		
Playback No. (PN)	γ	0.603	[0.272; 0.934]	<0.001
Random effect variance	u^2_k	1.299	[0.863; 1.957]	<0.001
Residual variance (intercept)	ε^2_{ik}	2.761	[1.423; 5.356]	0.003

[a] Reference.

4.5.2. Cognitive Performance

As in experiment 1, the performance data was first checked for the Stroop effect with a simple model considering congruency as the sole fixed effect. For both RT and ER a highly significant effect of congruency was given ($p < 0.001$), due to prolonged RTs and higher ERs during incongruent compared to congruent trials. Overall, incongruent trials (mean RT = 722 ms, SD = 119 ms) were answered 31 ms (or 5%) slower than congruent trials (mean RT = 691 ms, SD = 114 ms), and more errors were made in incongruent (mean ER = 2.0%, SD = 2.2%) than in congruent trials (mean ER = 1.3%, SD = 1.9%). Consequently, the effects on RTs and ERs were analysed separately for congruent and incongruent trials.

RT: Figure 11 shows the association of RTs with QTD, separately for congruent and incongruent trials in the Stroop task. RTs were linked to QTD, being longer for the longer (3-min) irregular breaks than the shorter (1-min) irregular and the regular breaks. This contrasts experiment 1, where the RTs were longer for the regular than the irregular (1-min) breaks (Figure 7). Besides, RTs decreased with increasing playback number (not shown). Congruent and incongruent trials were again affected similarly strong.

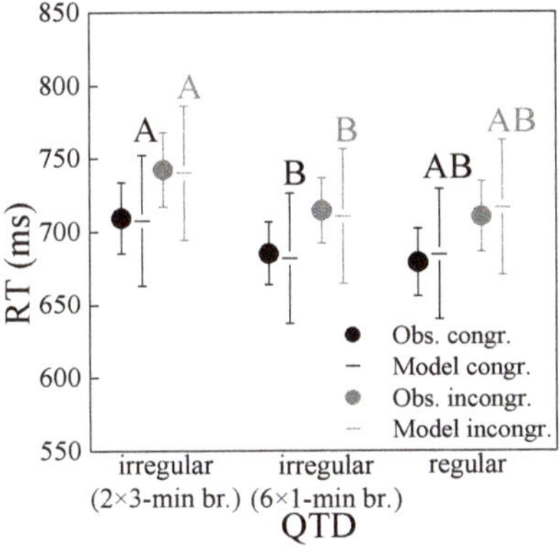

Figure 11. Reaction time (RT) as a function of quiet time distribution (QTD), for the congruent and incongruent trials of the Stroop task in experiment 2. Circles represent mean observed values (Obs.) with standard error bars, and horizontal lines the corresponding mixed-effects model with 95% confidence intervals. Significant differences between estimated marginal means ($p < 0.05$; pairwise comparisons with Bonferroni correction, performed separately for the congruent and incongruent Stroop tasks) of different QTDs are indicated by differing letters.

These observations and significant differences between long irregular and short irregular/regular QTD were confirmed by linear mixed-effects model analysis, which showed that RTs were significantly associated with QTD ($p < 0.02$) and playback number ($p < 0.001$) (details not shown). While the effect of QTD was significant, it was again small (around 30 ms compared to ~700 ms RTs on average, corresponding to a relative change of ~4%). RTs decreased with playback number increasing from 1–3 by some 100 and 90 ms for incongruent and congruent stimuli, respectively.

ER: In both congruent and incongruent trials, ER was neither associated with QTD nor playback number, which was also confirmed by mixed-effects model analysis (p's > 0.65 for QTD, p's > 0.05 for playback number).

5. Discussion

This study performed two unfocussed laboratory listening experiments to study how the macro-temporal pattern of different road traffic noise scenarios with rather low L_{Aeq} of ~45 dB(A) (experiment 1) and ~42 dB(A) (experiment 2), as might be expected in an office environment, affected short-term noise annoyance and cognitive performance in the Stroop task. A range of indicators for the macro-temporal pattern of the scenarios, including relative quiet time (RQT) and quiet time distribution (QTD), were quantified.

5.1. Annoyance

The experiments confirmed that quiet periods affect annoyance, revealing that annoyance ratings decreased with increasing RQT, at least up to some 60% (Figure 6). This is in line with literature [8–10,25,27,30,67]. Further, annoyance was linked with QTD. Shorter but more regular breaks were found to be perceived as less annoying than longer but irregular breaks of identical total duration. Similar insights as with RQT and QTD may also be obtained with the other indicators for the macro-temporal pattern (Table 2), which were closely related to either RQT or QTD (Table A1). For example, the number of events (negatively correlated with RQT) positively correlates with annoyance, which was also found for aircraft noise in [20], while *IR* (positively correlated with RQT) shows a negative correlation with annoyance, confirming the findings of [5]. In interpreting our results on *IR*, one should keep in mind that with the exception of the reference scenario S0, all scenarios were highly intermittent (cf. Figures 4 and 8), with *IR* values of 74% and more. Our findings suggest that, at the same RQT (with the same number of events), the clustering of car pass-by events after prolonged quiet times (irregular QTD), giving a more distinct temporal pattern, was more annoying to the participants than the shorter but regular events. Thus, to optimize QTD in order to minimize annoyance, providing a smooth traffic flow without too many interruptions, e.g., by reducing traffic lights, might be beneficial. In line with this thought, a laboratory study found that at high traffic densities, road traffic noise at a roundabout was perceived as less unpleasant than at crossroads with traffic lights [68]. RQT, in contrast, can only be optimized (meaning, increasing the breaks) through reduced the traffic volume (e.g., with traffic and parking restrictions and charges in cities), which also positively affects the L_{Aeq}.

The present results on QTD contrast the conclusions of previous studies that suggest a minimal duration of one [25] or three minutes [27–29] for a quiet period to be valuable with respect to annoyance, and of another laboratory study that did not find the duration of quiet periods to affect annoyance [67]. Thus, while breaks between events (i.e., having certain quiet periods, here: RQT) do seem beneficial, the link of the distribution of noise breaks with annoyance was less clear, and the necessity of a minimal duration of the noise breaks could not be confirmed. However, given the relatively low sound exposure in the experiments with an L_{Aeq} of ~42–45 dB(A), the effects were moderate only, changing annoyance by 1.4 units on the 11-point scale for a RQT increase from 0–81%, and 0.5–0.7 units for longer irregular compared to shorter quiet times (QTD).

Overall, the moderate association of annoyance with relatively low-level road traffic noise (L_{Aeq} of 42–45 dB(A)) is in line with a recent laboratory study that found the link between subjective disturbance and road traffic noise with an L_{Aeq} of 35–41 dB(A) to be quite weak [16].

5.2. Cognitive Performance

Compared to annoyance, the association of the macro-temporal pattern with cognitive performance in terms of RT and ER in the Stroop task was less clear. While RQT did not affect performance, QTD was slightly linked to RTs, but the results of experiments 1 and 2 were not clear-cut. In experiment 1, short regular breaks were found to be associated with longer RTs than short irregular breaks (Figure 7), but not in experiment 2. Here, long irregular breaks resulted in prolonged RTs (Figure 11). Yet in both experiments, the association of RTs with QTD, while significant, was weak, with small relative changes in RT of less than 5%. Further, no association of ER with the macro-temporal pattern of the noise scenarios was found. Similar results were also found in a preliminary listening experiment to this study [44], where road traffic noise neither affected RT nor ER.

This unsystematic effect pattern of the different noise scenarios on performance in the Stroop task might be due to their effect on attentional functions being comparatively smaller than their effect on noise annoyance, and because the applied experimental procedure did not allow for a more sensitive analysis of performance data. That is to say, the road traffic noise scenarios used in this experiment may have had too few salient changes (deviants) in terms of transitions from noisy to quiet periods (and back) diverting the attentional focus away from the task at hand to measure an effect on performance in the Stroop task when considering all trials worked out. However, the analysis of performance data could not be limited to those trials of the Stroop task that were performed at the time of, or shortly after, the salient changes in the road traffic noise scenarios. This was because the processing of the Stroop trials was self-paced in the present experiments, so that the relevant individual trials in the cognitive task could not be identified. In contrast, the above-mentioned laboratory study [16] found transitional phases in road traffic noise scenarios to affect reading task performance. Reading speed decreased as the sound level increased (rising front of an event) and increased again during the descending front.

Nevertheless, the typical Stroop effect was found in both experiments. That is, RTs were prolonged and ERs were increased for incongruent items, in which two dimensions of the visual stimulus did not match, compared to congruent items. This indicates that the participants seriously worked on the given cognitive task, and that our study in fact comprised unfocused listening experiments to investigate annoyance. Since performance in the Stroop task versions used here hardly changed during the different road traffic noise scenarios and, moreover, did not change systematically between the two experiments, differences in annoyance ratings can be assumed to not be moderated or even caused by performance effects (i.e., one was not annoyed because he/she could not perform well). Instead, the observed annoyance effects can be indeed attributed to the differing macro-temporal pattern of road traffic noise. In that context, it would be interesting to study the effects on noise annoyance in situations where also performance in (possible more difficult) cognitive tasks is affected by the macro-temporal pattern of road traffic noise.

5.3. Strengths and Limitations

A particular asset of the current study is that both, noise annoyance and cognitive performance, were mutually studied in two experiments to evaluate potential effects of road traffic noise comprehensively. While similar studies are available for background speech and music [39–41], studies involving road traffic noise to investigate such mutual effects are rare [16,17]. Besides, our design revealed that the associations of annoyance and performance with the acoustic characteristics (RQT or QTD) are quite different.

The study also faces certain limitations. As is generally true for laboratory studies, the ecological validity is limited due to the laboratory setting and the rather limited number of participants. Further, inferring from short-term noise annoyance in the laboratory to long-term annoyance in the field still needs to be verified ([69]), and inferring from cognitive performance tasks to long-term performance in office environments is similarly challenging.

Also some specific limitations apply. Above all, adopting the design to allow for a more sensitive analysis of performance data, specifically aiming at the transitional phases between quiet and loud periods (see above), would be beneficial. Besides, varying the L_{Aeq}, which is a decisive factor for road traffic noise annoyance (e.g., [45,68]) would add an important dimension to the outcomes. If the L_{Aeq} was sufficiently high to substantially affect cognitive performance, one could also study the effect of reduced performance on (noise) annoyance. These limitations could be addressed and improved in future studies (cf. Section 5.4).

5.4. Outlook

Our experiment revealed that, for moderate sound exposure in an office environment, the macro-temporal pattern of road traffic noise affects annoyance. This was true although participants were not actively listening to the noise but were working on a cognitive task, and even though performance on that task was not systematically affected by the noise. Future research might test whether the association of the macro-temporal pattern of the road traffic noise scenarios with annoyance is different if participants actively listen to them (e.g., during relaxation in a mock garden environment). This could be studied in a focussed listening experiment, where only the sound to be subjectively evaluated is presented, without any cognitive task to be performed.

Besides, follow-up experiments focusing more on the effects of road traffic noise scenarios on attentional functions might be set-up in such a way that the relevant trials in the cognitive task at the time of, or shortly after, the salient changes in the noise scenarios can be identified (i.e., non-self-paced trials or event based data logging). Then one could test more sensitively than in our experiments whether the transitions from traffic noise to quiet periods and back, and/or irregular breaks as unanticipated changes in the auditory background cause attentional capture.

In the experiments presented here, the levels were as one might well find them in an office environment. However, people are also exposed to traffic noise in street cafés, on balconies and in front gardens, where the sound levels can be significantly higher. Also there, people spend longer time periods and concentrate on certain cognitive tasks, if they have to or wish to. Consequently, further unfocussed listening experiments similar to the experiments presented here would be desirable to study the effect of macro-temporal pattern on annoyance and cognitive performance under substantially higher sound exposure (e.g., L_{Aeq} = 55–60 dB(A)). Such experiments could help further filling the gap in knowledge on the links between annoyance, performance and macro-temporal pattern of environmental sounds.

6. Conclusions

In unfocussed laboratory listening experiments, the associations of annoyance and cognitive performance with the macro-temporal pattern of relatively low-level road traffic noise situations were investigated in a mock office environment. In line with literature, annoyance decreased with increasing total duration of quiet periods. Also the distribution of the quiet times affected annoyance. Shorter but more regular breaks were found to be less annoying than longer but irregular breaks of identical total duration; a minimal necessary duration of noise breaks as proposed in literature could thus not be confirmed. Cognitive performance in an attention-based task, in contrast, did not systematically vary with the macro-temporal pattern of the situations. Thus, while the macro-temporal pattern of road traffic noise situations with moderate sound exposure seems playing a minor role for cognitive performance, it may still be important for annoyance of office staff.

Author Contributions: Conceptualization, A.T., S.J.S., J.M.W. and B.S.; methodology, A.T. and S.J.S.; experimental setup and conduct, L.B. and A.T.; data analysis, A.T. and B.S.; investigation, B.S., S.J.S. and A.T.; writing–original draft preparation, B.S., A.T., S.J.S., J.M.W., M.B. and L.B.; project administration, A.T. and J.M.W.; funding acquisition, A.T. and J.M.W. All authors have read and agreed to the published version of the manuscript.

Funding: This study (research project SiLENTIUM) was funded by the Swiss Federal Office for the Environment FOEN (assignment No. 5211.01520.100/contract No. 16.0076.PJ/R111-2085). The contribution by S.J.S. was supported by a grant from the HEAD-Genuit-Stiftung (HEAD-Genuit-Foundation: P-16/10-W).

Institutional Review Board Statement: The study was conducted in accordance with the Declaration of Helsinki, and approved by the ethics committee of Empa (approval CMI 2019-224 of 30 October 2019).

Informed Consent Statement: Informed consent was obtained from all participants involved in the study.

Data Availability Statement: Not applicable.

Acknowledgments: The authors are grateful to the participants of the laboratory experiments. They would further like to thank Markus Haselbach for his help in preparing the experimental laboratory setup.

Conflicts of Interest: The authors declare no conflict of interest. Mark Brink works for the funding agency FOEN, but contributed to this study in a purely scientific way and at the request of the other co-authors.

Appendix A

Table A1. Correlation analysis: Scatterplots and Spearman's rank correlation coefficient (r_s) [60] of indicators for the macro-temporal pattern of the road traffic noise scenarios S0–S6 of experiment 1.

Indicators	N	RQT (%)	IR (%)	CMT	$L_{AF,max}$ (dB(A))
N		−1.000 **	−0.973 **	−0.120	−0.911 **
RQT (%)			0.973 **	0.120	0.911 **
IR (%)				0.371	0.906 **
CMT					0.270
$L_{AF,max}$ (dB(A))					

** $p < 0.01$.

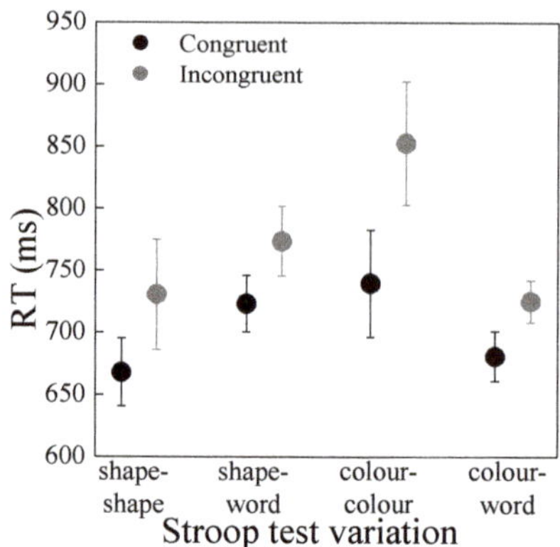

Figure A1. Results of a pilot experiment to this study (details see [44]): Mean reaction time (RT) with standard error bars, shown separately for congruent and incongruent trials, for four variations of the Stroop task: (i) shape test naming the shape of a geometric form with a word written within (shape-shape), (ii) shape test naming the written word within a geometric form instead of its form (shape-word), (iii) colour test naming the print colour instead of the semantics of the word (colour-colour), (iv) colour test naming the semantics instead of the print colour of the word (colour-word).

References

1. WHO. *Burden of Disease from Environmental Noise. Quantification of Healthy Life Years Lost in Europe*; World Health Organization (WHO), Regional Office for Europe: Copenhagen, Denmark, 2011. Available online: http://www.euro.who.int/en/publications (accessed on 10 February 2022).
2. Guski, R.; Schreckenberg, D.; Schuemer, R. WHO environmental noise guidelines for the European region: A systematic review on environmental noise and annoyance. *Int. J. Environ. Res. Public Health* **2017**, *14*, 1539. [CrossRef] [PubMed]
3. ISO. *ISO/TS 15666*; Technical Specification: Acoustics–Assessment of Noise Annoyance by Means of Social and Socio-Acoustic Surveys. 2nd ed. 2021-05. International Organisation for Standardization (ISO): Geneva, Switzerland, 2021.
4. Miedema, H.; Oudshoorn, C. Annoyance from transportation noise: Relationships with exposure metrics DNL and DENL and their confidence intervals. *Environ. Health Perspect.* **2001**, *109*, 409–416. [CrossRef] [PubMed]
5. Brink, M.; Schäffer, B.; Vienneau, D.; Foraster, M.; Pieren, R.; Eze, I.C.; Cajochen, C.; Probst-Hensch, N.; Röösli, M.; Wunderli, J.-M. A survey on exposure-response relationships for road, rail, and aircraft noise annoyance: Differences between continuous and intermittent noise. *Environ. Int.* **2019**, *125*, 277–290. [CrossRef] [PubMed]
6. WHO. *Environmental Noise Guidelines for the European Region*; World Health Organization (WHO), Regional Office for Europe: Copenhagen, Denmark, 2018. Available online: http://www.euro.who.int/en/health-topics/environment-and-health/noise/environmental-noise-guidelines-for-the-european-region (accessed on 10 February 2022).
7. Fleischer, G. Vorschlag für die Bewertung von Lärm und Ruhe. *Kampf Dem Lärm* **1979**, *26*, 129–134.
8. Estévez-Mauriz, L.; Forssén, J. Dynamic traffic noise assessment tool: A comparative study between a roundabout and a signalised intersection. *Appl. Acoust.* **2018**, *130*, 71–86. [CrossRef]
9. Taghipour, A.; Pieren, R.; Schäffer, B. Relative duration of quiet periods between events influences noise annoyance: A laboratory experiment with helicopter sounds. In Proceedings of the 23rd International Congress on Acoustics (ICA), Aachen, Germany, 9–13 September 2019; pp. 8011–8018.
10. Fleischer, G. Argumente für die Berücksichtigung der Ruhe in der Lärmbekämpfung. *Kampf Dem Lärm* **1978**, *25*, 69–74.
11. Wunderli, J.M.; Pieren, R.; Habermacher, M.; Vienneau, D.; Cajochen, C.; Probst-Hensch, N.; Röösli, M.; Brink, M. Intermittency ratio: A metric reflecting short-term temporal variations of transportation noise exposure. *J. Expo. Sci. Environ. Epidemiol.* **2015**, *26*, 575–585. [CrossRef]
12. Clark, C.; Paunovic, K. WHO environmental noise guidelines for the European region: A systematic review on environmental noise and cognition. *Int. J. Environ. Res. Public Health* **2018**, *15*, 285. [CrossRef]
13. Aletta, F.; Oberman, T.; Kang, J. Associations between positive health-related effects and soundscapes perceptual constructs: A systematic review. *Int. J. Environ. Res. Public Health* **2018**, *15*, 2392. [CrossRef]
14. Worldometer Webpage: Worldometer. Available online: https://www.worldometers.info (accessed on 10 February 2022).
15. EEA. *Environmental Noise in Europe—2020*; EEA Report No 22/2019; European Environment Agency (EEA): Copenhagen, Denmark, 2020. Available online: https://www.eea.europa.eu/publications/environmental-noise-in-europe (accessed on 10 February 2022).

16. Lavandier, C.; Regragui, M.; Dedieu, R.; Royer, C.; Can, A. Influence of road traffic noise peaks on reading task performance and disturbance in a laboratory context. *Acta Acust.* **2022**, *6*, 3. [CrossRef]

17. Schlittmeier, S.J.; Feil, A.; Liebl, A.; Hellbrück, J. The impact of road traffic noise on cognitive performance in attention-based tasks depends on noise level even within moderate-level ranges. *Noise Health* **2015**, *17*, 148–157. [CrossRef] [PubMed]

18. ECAC. *ECAC.CEAC DOC.29: Report on Standard Method of Computing Noise Contours around Civil Airports, Volume 2: Technical Guide*, 4th ed.; European Civil Aviation Conference (ECAC): Neuilly-sur-Seine, France, 2016. Available online: https://www.ecac-ceac.org/images/documents/ECAC-Doc_29_4th_edition_Dec_2016_Volume_2.pdf (accessed on 10 February 2022).

19. Gjestland, T.; Gelderblom, F.B. Prevalence of noise induced annoyance and its dependency on number of aircraft movements. *Genome* **2017**, *103*, 28–33. [CrossRef]

20. Haubrich, J.; Benz, S.; Brink, M.; Guski, R.; Isermann, U.; Schäffer, B.; Schmid, R.; Schreckenberg, D.; Wunderli, J.M. Leq + X: Re-Assessment of exposure-response relationships for aircraft noise annoyance and disturbance to improve explained variance. In Proceedings of the 23rd International Congress on Acoustics (ICA), Aachen, Germany, 9–13 September 2019; pp. 1523–1530.

21. Taghipour, A.; Pieren, R.; Schäffer, B. Short-term annoyance reactions to civil helicopter and propeller-driven aircraft noise: A laboratory experiment. *J. Acoust. Soc. Am.* **2019**, *145*, 956–967. [CrossRef] [PubMed]

22. Can, A.; Gauvreau, B. Describing and classifying urban sound environments with a relevant set of physical indicators. *J. Acoust. Soc. Am.* **2015**, *137*, 208–218. [CrossRef]

23. Asensio, C.; Aumond, P.; Can, A.; Gascó, L.; Lercher, P.; Wunderli, J.M.; Lavandier, C.; de Arcas, G.; Ribeiro, C.; Muñoz, P.; et al. A taxonomy proposal for the assessment of the changes in soundscape resulting from the COVID-19 lockdown. *Int. J. Environ. Res. Public Health* **2020**, *17*, 4205. [CrossRef]

24. Bockstael, A.; De Coensel, B.; Lercher, P.; Botteldooren, D. Influence of temporal structure of the sonic environment on annoyance. In Proceedings of the 10th International Congress on Noise as a Public Health Problem (ICBEN), London, UK, 24–28 July 2011; pp. 945–952.

25. Finke, H.O. Messung und Beurteilung der "Ruhigkeit" bei Geräuschimmissionen. *Acoustica* **1980**, *46*, 141–148.

26. Krause, M. Messung der Ruhe. *Kampf Dem Lärm* **1979**, *25*, 75–79.

27. Guski, R. Können Ruhepausen im Lärm wahrgenommen werden? *Z. Lärmbek.* **1988**, *35*, 69–73.

28. Guski, R. First steps toward the concept of quietness and its psychological and acoustical determinants. In Proceedings of the International Conference on Noise Control Engineering (Inter-Noise), Edinburgh, UK, 13–15 July 1983; pp. 843–846.

29. Guski, R. Is there any need for quiet periods in discontinuous noise? In Proceedings of the International Conference on Noise Engineering (Inter-Noise), Munich, Germany, 18–20 September 1985; pp. 985–988.

30. Dornic, S.; Laaksonen, T. Continuous noise, intermittent noise, and annoyance. *Percept. Mot. Ski.* **1989**, *68*, 11–18. [CrossRef]

31. Belojević, G. Noise and performance: Research in Central, Eastern and South-Eastern Europe and Newly Independent States. *Noise Health* **2013**, *15*, 2–5. [CrossRef]

32. Szalma, J.L.; Hancock, P.A. Noise effects on human performance: A meta-analytic synthesis. *Psychol. Bull.* **2011**, *137*, 682–707. [CrossRef]

33. Smith, A. An update on noise and performance: Comment on Szalma and Hancock (2011). *Psychol. Bull.* **2012**, *138*, 1262–1268. [CrossRef] [PubMed]

34. Belojević, G.; Öhrström, E.; Rylander, R. Effects of noise on mental performance with regard to subjective noise sensitivity. *Int. Arch. Occup. Environ. Health* **1992**, *64*, 293–301. [CrossRef] [PubMed]

35. Enmarker, I. The effects of meaningful irrelevant speech and road traffic noise on teachers' attention, episodic and semantic memory. *Scand. J. Psychol.* **2004**, *45*, 393–405. [CrossRef] [PubMed]

36. Hygge, S.; Boman, E.; Enmarker, I. The effects of road traffic noise and meaningful irrelevant speech on different memory systems. *Scand. J. Psychol* **2003**, *44*, 13–21. [CrossRef] [PubMed]

37. Lee, P.J.; Jeon, J.Y. Relating traffic, construction, and ventilation noise to cognitive performances and subjective perceptions. *J. Acoust. Soc. Am.* **2013**, *134*, 2765–2772. [CrossRef]

38. Hughes, R.W.; Vachon, F.; Jones, D.M. Auditory attentional capture during serial recall: Violations at encoding of an algorithm-based neural model? *J. Exp. Psychol. Learn. Mem. Cogn.* **2005**, *31*, 736–749. [CrossRef]

39. Schlittmeier, S.J.; Hellbrück, J.; Thaden, R.; Vorländer, M. The impact of background speech varying in intelligibility: Effects on cognitive performance and perceived disturbance. *Ergonomics* **2008**, *51*, 719–736. [CrossRef]

40. Schlittmeier, S.J.; Hellbrück, J. Background music as noise abatement in open-plan offices: A laboratory study on performance effects and subjective preferences. *Appl. Cogn. Psychol.* **2009**, *23*, 684–697. [CrossRef]

41. Schlittmeier, S.J.; Liebl, A. The effects of intelligible irrelevant background speech in offices–Cognitive disturbance, annoyance, and solutions. *Facilities* **2015**, *33*, 61–75. [CrossRef]

42. Stroop, J.R. Studies of interference in serial verbal reactions. *J. Exp. Psychol.* **1935**, *18*, 643–662. [CrossRef]

43. Gier, V.S.; Kreiner, D.S.; Solso, R.L.; Cox, S.L. The hemispheric lateralization for processing geometric word/shape combinations: The Stroop-shape effect. *J. Gen. Psychol.* **2010**, *137*, 1–19. [CrossRef] [PubMed]

44. Taghipour, A.; Bartha, L.; Schlittmeier, S.J.; Schäffer, B. Effects of noise on performance and perceived annoyance in Stroop tasks. In Proceedings of the e-Forum Acusticum 2020, Lyon, France, 7–11 December 2020; pp. 63–69.

45. Schäffer, B.; Schlittmeier, S.J.; Pieren, R.; Heutschi, K.; Brink, M.; Graf, R.; Hellbrück, J. Short-term annoyance reactions to stationary and time-varying wind turbine and road traffic noise: A laboratory study. *J. Acoust. Soc. Am.* **2016**, *139*, 2949–2963. [CrossRef] [PubMed]

46. Nilsson, M.E. A-weighted sound pressure level as an indicator of short-term loudness or annoyance of road-traffic sound. *J. Sound Vib.* **2007**, *302*, 197–207. [CrossRef]

47. Fastl, H.; Zwicker, E. *Psychoacoustics: Facts and Models*, 3rd ed.; Springer: Berlin/Heidelberg, Germany, 2007.

48. Bolin, K.; Bluhm, G.; Nilsson, M.E. Listening test comparing A-weighted and C-weighted sound pressure level as indicator of wind turbine noise annoyance. *Acta Acust. United Acust.* **2014**, *100*, 842–847. [CrossRef]

49. Alamir, M.A.; Hansen, K.L.; Zajamsek, B.; Catcheside, P. Subjective responses to wind farm noise: A review of laboratory listening test methods. *Renew. Sustain. Energy Rev.* **2019**, *114*, 109317. [CrossRef]

50. Van Renterghem, T.; Bockstael, A.; De Weirt, V.; Botteldooren, D. Annoyance, detection and recognition of wind turbine noise. *Sci. Total Environ.* **2013**, *456–457*, 333–345. [CrossRef] [PubMed]

51. LSV. Lärmschutz-Verordnung (LSV) vom 15. Dezember 1986 (Stand am 1. Juli 2021). SR 814.41. 1986. Available online: https://www.fedlex.admin.ch/eli/cc/1987/338_338_338/de (accessed on 10 February 2022).

52. Basner, M.; Isermann, U.; Samel, A. Die Umsetzung der DLR-Studie in einer lärmmedizinischen Beurteilung für ein Nachtschutzkonzept. *Z. Lärmbekämpf.* **2005**, *52*, 109–123.

53. Locher, B.; Piquerez, A.; Habermacher, M.; Ragettli, M.; Röösli, M.; Brink, M.; Cajochen, C.; Vienneau, D.; Foraster, M.; Müller, U.; et al. Differences between outdoor and indoor sound levels for open, tilted, and closed windows. *Int. J. Environ. Res. Public Health* **2018**, *15*, 149. [CrossRef]

54. Brink, M.; Schäffer, B.; Pieren, R.; Wunderli, J.M. Conversion between noise exposure indicators Leq_{24h}, L_{Day}, $L_{Evening}$, L_{Night}, L_{dn} and L_{den}: Principles and practical guidance. *Int. J. Hyg. Environ. Health* **2018**, *221*, 54–63. [CrossRef]

55. Nordtest. *Acoustics: Human Sound Perception–Guidelines for Listening Tests. Nordtest Method, NT ACOU 111, Approved 2002-05*; Nordtest: Espoo, Finland, 2002. Available online: http://www.nordtest.info/wp/2002/05/01/acoustics-human-sound-perception-guidelines-for-listening-tests-nt-acou-111/ (accessed on 10 February 2022).

56. Ellermeier, W.; Hellbrück, J.; Kohlrausch, A.; Zeitler, A. *Kompendium zur Durchführung von Hörversuchen in Wissenschaft und industrieller Praxis*; Deutsche Gesellschaft für Akustik e.V.: Berlin, Deutschland, 2008. Available online: https://www.dega-akustik.de/fileadmin/dega-akustik.de/publikationen/Kompendium_Hoerversuche_2008.pdf (accessed on 10 February 2022).

57. Jensen, A.R.; Rohwer, W.D. The Stroop color-word test—A review. *Acta Psychol.* **1966**, *25*, 36–93. [CrossRef]

58. MacLeod, C.M. Half a century of research on the Stroop effect: An integrative review. *Psychol. Bull.* **1991**, *109*, 163–203. [CrossRef] [PubMed]

59. Peirce, J.; Gray, J.R.; Simpson, S.; MacAskill, M.; Höchenberger, R.; Sogo, H.; Kastman, E.; Lindeløv, J.K. PsychoPy2: Experiments in behavior made easy. *Behav. Res. Methods* **2019**, *51*, 195–203. [CrossRef] [PubMed]

60. Spearman, C. The proof and measurement of association between two things. *Am. J. Psychol.* **1904**, *15*, 72–101. [CrossRef]

61. Fields, J.M.; De Jong, R.G.; Gjestland, T.; Flindell, I.H.; Job, R.F.S.; Kurra, S.; Lercher, P.; Vallet, M.; Yano, T.; Guski, R.; et al. Standardized general-purpose noise reaction questions for community noise surveys: Research and a recommendation. *J. Sound Vib.* **2001**, *242*, 641–679. [CrossRef]

62. Johnson, D.R.; Creech, J.C. Ordinal measures in multiple indicator models: A simulation study of categorization error. *Am. Sociol. Rev.* **1983**, *48*, 398–407. [CrossRef]

63. Norman, G. Likert scales, levels of measurement and the "laws" of statistics. *Adv. Health Sci. Educ.* **2010**, *15*, 625–632. [CrossRef]

64. Sullivan, G.M.; Artino, A.R., Jr. Analyzing and interpreting data from Likert-type scales. *J. Grad. Med. Educ.* **2013**, *5*, 541–542. [CrossRef]

65. Pinheiro, J.C.; Bates, D.M. *Mixed-Effects Models in S and S-Plus*, 1st ed.; Springer: New York, NY, USA, 2000; p. 528.

66. Cohen, B.H. *Explaining Psychological Statistics*, 4th ed.; John Wiley and Sons, Inc.: Hoboken, NJ, USA, 2013.

67. Gille, L.A.; Marquis-Favre, C.; Klein, A. Noise annoyance due to urban road traffic with powered-two-wheelers: Quiet periods, order and number of vehicles. *Acta Acust. United Acust.* **2016**, *102*, 474–487. [CrossRef]

68. Trollé, A.; Terroir, J.; Lavandier, C.; Marquis-Favre, C.; Lavandier, M. Impact of urban road traffic on sound unpleasantness: A comparison of traffic scenarios at crossroads. *Appl. Acoust.* **2015**, *94*, 46–52. [CrossRef]

69. Guski, R.; Bosshardt, H.-G. Gibt es eine "unbeeinflußte" Lästigkeit? *Z. Lärmbekämpf.* **1992**, *39*, 67–74.

International Journal of
***Environmental Research
and Public Health***

Article

A Noise Control Method Using Adaptive Adjustable Parametric Array Loudspeaker to Eliminate Environmental Noise in Real Time

Yinsheng Li and Wei Zheng *

Key Laboratory for Optoelectronic Technology and System of the Education Ministry of China, College of Optoelectronic Engineering, Chongqing University, Chongqing 400044, China; yinsheng.li@cqu.edu.cn
* Correspondence: zw3475@cqu.edu.cn

Abstract: Long-term exposure to environmental noise is dangerous to human health. Therefore, there is an urgent need to suppress or eliminate environmental noise. Due to the limitation of environmental space, the use of reverse sound waves emitted by loudspeakers for noise elimination has been widely used in noise control. However, because of the omni-directionality of sound propagation, a traditional voice coil loudspeaker (VCL) is used as a secondary source (emission reverse sound wave). It is easy to increase the sound pressure in non-target areas and form significant acoustic feedback to the reference source. Therefore, we propose an online secondary path modeling method using an adjustable parametric array loudspeaker (PAL) based on ultrasounds to eliminate environmental noise in real time. According to the different distance of the target, the size of the PAL is adjusted adaptively to realize the noise control of different long-distance targets. The distribution of quiet areas is discussed. The experimental results showed that a PAL as a secondary source had the same noise reduction effect as a traditional VCL, but it had longer propagation distance, smaller sound feedback and a more regular and controllable distribution of quiet areas. These research findings have great potential for improving environmental noise and creating a quiet environment.

Keywords: ultrasound; environmental noise; active noise control; adjustable PAL; quiet areas

Citation: Li, Y.; Zheng, W. A Noise Control Method Using Adaptive Adjustable Parametric Array Loudspeaker to Eliminate Environmental Noise in Real Time. *Int. J. Environ. Res. Public Health* **2022**, *19*, 269. https://doi.org/10.3390/ijerph19010269

Academic Editors: Gaetano Licitra, Luca Fredianelli and Peter Lercher

Received: 2 December 2021
Accepted: 23 December 2021
Published: 27 December 2021

Publisher's Note: MDPI stays neutral with regard to jurisdictional claims in published maps and institutional affiliations.

1. Introduction

Environmental noise is everywhere. However, when people work and live in a high-noise environment for a long time, it brings great risks to people's health, such as hearing loss and noise trouble [1,2]. As people mainly work and live in indoor environments, the application scenarios of noise control are also mainly aimed at indoor environmental noise. At present, environmental noise control methods such as sound absorption and vibration isolation are mainly used. These methods have a good effect on reducing medium and high frequency noise in space. However, the passive noise control equipment is complex, and the low-frequency control effect is poor, which is greatly restrictive. In recent years, active noise control (ANC) can effectively make up for the deficiencies of passive noise control, and has become a hot research topic in the field of noise control [3,4]. With the development of adaptive active control technology, active noise control has been widely used in offices, automobiles and other spaces [5,6].

Active noise control, through the adaptive filtering algorithm, controls the secondary source to produce a secondary sound field, which is superimposed with the primary sound field, so as to achieve the purpose of noise suppression [7]. Adaptive filtering algorithm is the focus of active noise control. The well-known filtered-x least mean square (FxLMS) algorithm was proposed by Burgess in 1981 [8]. The FxLMS algorithm is still the most widely used because of its simplicity and efficiency. The FxLMS algorithm uses the secondary path response function to filter the reference signal, so the secondary path needs to be fitted. The secondary path fitting methods include online fitting and offline fitting.

In practical use, the secondary path response function is changed due to the change of environment and device aging, so it is necessary to estimate the secondary path online. Eriksson et al., used the method of injecting auxiliary noise to fit the secondary path online for the first time [9]. In recent years, many researchers have used different strategies to improve the convergence speed and steady-state performance of the online secondary path modeling [10–13].

Active noise control uses traditional voice coil loudspeakers (VCLs) as the secondary source to suppress the noise in the target areas. Due to the omni-directivity of sound propagation, noise reduction is achieved at the target point, and the overflow of reverse sound waves increases the sound pressure level (SPL) of other adjacent areas [14,15]. When a multi-channel ANC system is used, the noise acquisition of the error microphone between the secondary sources produces crosstalk, which increases the additional calculation cost in the control process [16,17]. Due to the omni-directivity of sound propagation in traditional voice coil loudspeakers, it is easy to increase the acoustic feedback to the reference signal [18,19]. Therefore, there are still many shortcomings to using a traditional voice coil loudspeaker as a secondary source for noise control.

Ultrasound is a sound wave with a frequency higher than 20 kHz. It is used in parametric array loudspeakers by using the directivity of ultrasound. Parametric array loudspeakers (PALs) use the nonlinear effect of ultrasound wave in the medium to produce audible sound. Compared with traditional voice coil loudspeakers, parametric array loudspeakers, at both low frequency and high frequency, have significant high directivity [20,21]. Therefore, PALs can produce audible sound in the target areas without interfering with other areas [22]. In active noise control, researchers have used a PAL as the secondary source, which verified its feasibility [23–25]. More and more successful attempts have been made to use a PAL as a secondary source. Ganguly et al., used one as the secondary source to control the noise of the 1.06 m target [26]. Tanaka et al., used PALs to establish a silent area for the left and right ear areas of workers at 1.5 m [27]. However, in the above applications, the distance between the secondary source PAL and the noise reduction target is fixed and placed at a short distance, so it is impossible to adjust the distance of different targets adaptively. In addition, compared with traditional voice coil loudspeakers, the research on the many unique characteristics of PALs is still limited, and further research is still needed.

In previous studies using a PAL as a secondary source, its size and power were fixed. Because the spatial distance between the secondary source and the noise reduction target is fixed, once the target position is changed, new problems arise. When the target is too close to the secondary source, this can cause sound field overflow and multiple harmonics to form superposition and interference with the other adjacent sound fields. When it is too far, the noise reduction performance is greatly reduced. In view of the above research status, this paper uses the adjustable parametric array loudspeaker as the secondary source. According to the target distance, we adaptively adjusted the size and power of a PAL to achieve noise control in the target area. In addition, the size of acoustic feedback and the distribution of quiet areas between PALs and traditional VCLs are discussed and compared. Combined with the characteristics of PALs, it is helpful to improve the performance of ANC systems.

The rest of this paper is organized as follows. Section 2 presents the active noise control system based on online secondary path modeling and the theory of differential frequency sound field. In Section 3, experiments are carried out based on adjustable PALs and compared with traditional VCLs, and the experimental results are given. The conclusion is given in Section 4.

2. Methods

2.1. Online Secondary Path Modeling

Due to its simplicity and effectiveness, FxLMS algorithm is the most widely used adaptive algorithm. In the adaptive FxLMS algorithm, we need to identify the secondary path transfer function to obtain its estimated value. The process of using adaptive filtering principle to estimate the secondary path is called secondary path modeling. Secondary path modeling is divided into offline estimation and online estimation. The offline estimation is applicable to the application scenarios where the secondary path is stable or changes slowly. In this paper, the secondary path between the secondary source and the error microphone changes with the noise reduction target position, and the corresponding secondary path response function S(z) also changes. Therefore, the secondary path can only be used for online estimation.

Online secondary path modeling is employed to estimate the transfer function of the secondary path in real time according to the changes of the actual environment in the process of active noise control. The power of auxiliary noise affects the convergence speed of the secondary path fitting filter. The larger the auxiliary noise power, the faster the convergence speed of the secondary path filter, but it decreases the noise reduction performance of the system. When the auxiliary noise is too low, the convergence speed is reduced, which affects the stability of the ANC filter [11,12]. In order to balance the contradiction between the two, researchers proposed a method of variable power auxiliary noise. In the initial stage of system operation or when the secondary path changes, high-power auxiliary noise is used to speed up the convergence speed. When the system is stable, low power auxiliary noise is used. The most widely used is the auxiliary noise power regulation strategy proposed by Akhtar and Carini [11,28]. On this basis, this article adopts a more sensitive auxiliary noise power adjustment method.

Figure 1 shows the online secondary path modeling with variable power auxiliary noise. Among them, $S(z)$ is the transfer function of the secondary path, and $\hat{S}(z)$ is the estimated quantity of the transfer function of the secondary path. The white noise generator generates a white noise signal $v(n)$ that is not related to the reference signal $x(n)$, and superimposes it with the output signal $y(n)$ generated by the ANC filter. After passing through the secondary path $S(z)$, $y'(n) - v'(n)$ is obtained, and superimposed with the desired signal $d(n)$, the error signal of the entire ANC system is

$$e(n) = d(n) - y'(n) + v'(n). \tag{1}$$

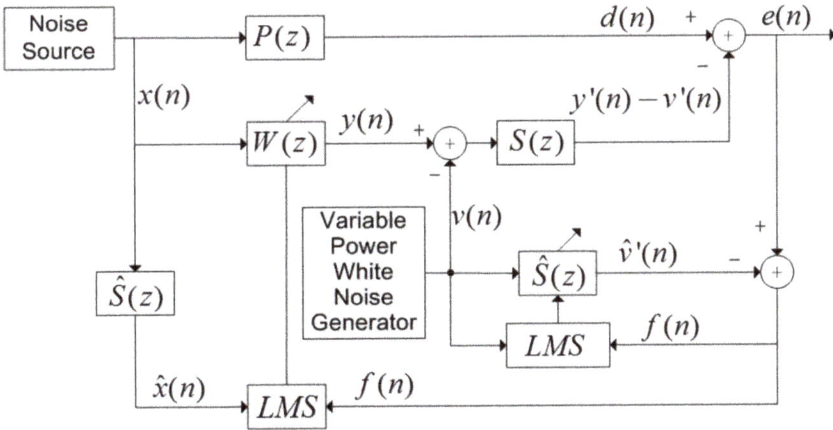

Figure 1. Online secondary path modeling with variable power auxiliary noise. $x(n)$ is the signal of noise source, $P(z)$ is the primary path transfer function, and the weight vector $W(z)$ is iteratively updated by LMS algorithm.

After the white noise signal $v(n)$ passes through the secondary pass filter, the estimated quantity $\hat{v}'(n)$ is obtained, and after superimposing it with the error signal $e(n)$, the error quantity $f(n)$ used to update the ANC filter coefficient and the secondary path filter coefficient is obtained:

$$\begin{aligned} f(n) &= e(n) - \hat{v}'(n) \\ &= d(n) - y'(n) + v'(n) - \hat{v}'(n) \\ &= e_x(n) + e_v(n) \end{aligned} \tag{2}$$

Among them, $e_x(n) = d(n) - y'(n)$, $e_v(n) = v'(n) - \hat{v}'(n)$. $e_x(n)$ represents the error caused by the ANC filter, and $e_v(n)$ represents the error caused by the secondary path filter. The ANC filter and secondary path filter coefficient update formula is:

$$\hat{S}(n+1) = \hat{S}(n) + \mu_s v(n)f(n), \tag{3}$$

$$W(n+1) = W(n) + \mu_w \hat{x}'(n)f(n). \tag{4}$$

Among them, μ_s and μ_w are the step factors of the secondary path filter and the ANC filter, respectively. The white noise signal input vector is denoted as $v^T(n) = [v(n), v(n-1), \cdots, v(n-L+1)]^T$.

The commonly used auxiliary noise $v(n)$ power adjustment method is:

$$v_g(n) = \sqrt{(1 - \rho(n))\sigma^2_{v_{\min}} + \rho(n)\sigma^2_{v_{\max}}} \quad \bullet \quad v(n) = G(n)v(n), \tag{5}$$

$$\rho(n) = \frac{p_f(n)}{p_e(n)} = \frac{P_{[d(n)-y'(n)]} + P_{[v'(n)-\hat{v}'(n)]}}{P_{[d(n)-y'(n)]} + P_{[v'(n)]}}, \tag{6}$$

$p_f(n)$ and $p_e(n)$ are the power of the error signal $f(n)$ and $e(n)$ respectively. $G(n)$ is the auxiliary noise gain, $v_g(n)$ is the adjusted auxiliary noise signal, $\sigma^2_{v_{\min}}$ and $\sigma^2_{v_{\max}}$ are the minimum and maximum noise power, respectively. When the reference noise power is too large, the power that needs to be controlled at the error microphone is much greater than the auxiliary noise power, which causes large fluctuations in the secondary path filter. Therefore, it is necessary to consider the influence of the reference noise power when performing auxiliary noise power control.

For this reason, the ratio of the residual error power of the ANC filter to the power of the auxiliary noise filtered by the secondary path can be a fixed value.

$$\frac{E((d(n) - y'(n))^2)}{E((v'_g(n))^2)} = R = constant. \tag{7}$$

When the gain $G(n)$ of the auxiliary noise changes slowly,

$$E((v'_g(n))^2) = G^2(n)\big\|s(n)\big\|_2 E(v(n)^2), \tag{8}$$

$\|s(n)\|$ is the Euclidean norm of the secondary path coefficient vector.

$$E(e(n)^2) = E((d(n) - y'(n))^2) + E((v'_g(n))^2). \tag{9}$$

According to Equations (7)–(9), the calculation formula of the auxiliary noise gain can be obtained as:

$$G(n) = \sqrt{\frac{P_e(n)}{(R+1)P_{\hat{s}}(n)}}. \tag{10}$$

Among them, $P_{\hat{s}}(n)$ also uses exponential smoothing to estimate:

$$P_{\hat{s}}(n) = \lambda P_{\hat{s}}(n-1) + (1-\lambda)\hat{s}^T(n)\hat{s}(n), \tag{11}$$

λ is a forgetting factor close to 1. Because the residual noise is also related to the power level of the reference noise to a certain extent, the power of the residual noise may not fully reflect the closeness of the ANC filter to the steady state. In order to make the auxiliary noise power more sensitive to the state of ANC filter, the ratio of $P_x(n)$ and $P_e(n)$ is used to estimate the degree of convergence of the system.

$$R(n) = \frac{P_x(n)}{P_e(n)}.$$

(12)

Compared with the fixed scale coefficient, the auxiliary noise power has smaller power in the steady state. The calculation formula of power gain is:

$$G(n) = \sqrt{\frac{P_e(n)}{(R(n)+1)P_{\hat{s}}(n)}}.$$

(13)

Therefore, the auxiliary noise power adjustment formula is:

$$v_g(n) = G(n)v(n) = \sqrt{\frac{P_e(n)}{(R(n)+1)P_{\hat{s}}(n)}} \bullet v(n).$$

(14)

The above power regulation method is more sensitive to the state of ANC filter, has large power in the initial stage of system operation, can attenuate rapidly when the system tends to be stable, and the auxiliary noise power is greatly reduced in the steady state.

The existing FxLMS algorithm, normalized FxLMS algorithm and the online modeling method with variable power auxiliary noise used in this paper are simulated, respectively. The iterative mean square error (MSE) results are shown in Figure 2. The filter order of secondary path modeling is 32, and the iteration step of secondary path modeling is 0.01. It can be seen from the figure that the online modeling method with power auxiliary noise has faster convergence speed and lower steady-state offset.

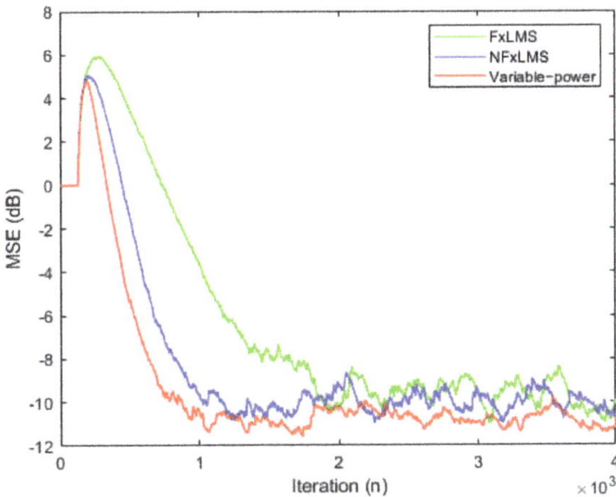

Figure 2. Performance comparison. The horizontal axis is iteration (n) and the vertical axis is MSE (dB). Legend is FxLMS, NFxLMS, and variable-power, respectively.

2.2. Theoretical Basis of Parametric Array

Parametric array loudspeaker is a kind of parametric loudspeaker which can control the audible sound in a specific direction. Using the nonlinear effect of ultrasound wave in air, an audible sound field with high directivity is formed in air. The key of parametric array theory model is the establishment of nonlinear wave equation when sound wave propagates in air. Khokhlov, Zabolotskaya and Kuznetsov fully considered the absorption, scattering and nonlinear effects of finite amplitude sound beam in fluid and solid, and deduced the nonlinear equation of wave propagation in medium, which is the famous KZK equation [29,30]. The equation accurately describes the nonlinear propagation effect of sound wave in medium.

The expression of KZK equation is:

$$\frac{\partial^2 p}{\partial z \partial \tau} = \frac{c_0}{2} \nabla_\perp^2 p + \frac{\delta}{2c_0^3} \frac{\partial^3 p}{\partial \tau^3} + \frac{\beta}{2\rho_0 c_0^3} \frac{\partial^2 p^2}{\partial \tau^2},$$

(15)

where, p is the sound pressure, z is the propagation distance along the sound beam axis, c_0 is the sound velocity, $\tau = t - z/c_0$ is the delay time, δ is the sound scattering degree, β is the nonlinear coefficient, ρ_0 is the air density, and $\nabla_\perp^2$ is the XY plane Laplace operator perpendicular to the Z axis.

Because KZK equation involves many nonlinear parameters, such as scattering in the medium, heat propagation loss and molecular relaxation loss, it is difficult to obtain its exact analytical solution. The approximate solution or numerical solution is usually obtained by quasilinear method. It is assumed that the harmonic frequency component generated by the sound wave in the process of propagation has the fundamental frequency. Ignoring the third harmonic with small harmonic amplitude, it is only approximate to the second harmonic. At this time, the solution of KZK equation is:

$$p = p_1 + p_2,$$

(16)

where, p_1 is the primary wave sound pressure generated by sound wave, and p_2 is the second harmonic generated by nonlinear propagation, and its amplitude is less than p_1.

Introducing complex pressure amplitude q_n and set:

$$p_n(r, z, \tau) = \frac{1}{2j} q_n(r, z) e^{jn\omega\tau} + c.c. \qquad n = 1, 2,$$

(17)

where, $c.c.$ is the complex conjugate of the former term, and r is the distance between the projection of the midpoint of the sound field on the plane of the parametric array and the center of the plane. By applying Green's function and Hankel's transformation, the quasilinear solution of KZK equation is finally obtained as follows [31]:

$$q_1(r, z) = 2\pi \int_0^\infty q_1(r', 0) G_1(r, z|r', 0) r' dr',$$

(18)

$$q_2(r, z) = \frac{\pi \beta k}{\rho_0 c_0^2} \int_0^z \int_0^\infty q_1^2(r', z') G_2(r, z|r', z') r' dr' dz',$$

(19)

where, $k = \omega/c_0$ is the wave number, $G_n(r, z|r', z')$ is the Green function, $q_1(r, z)$ is the complex value of linear sound pressure of primary wave, and $q_2(r, z)$ is the complex value of second harmonic sound pressure caused by nonlinear propagation effect of primary wave. There are multiple integrals in the above formula, which is still not easy to calculate. By applying Hankel's transformation to the wave equation, Liauh et al., obtained the calculation formula of general circular piston source parameter array parameters ϕ [32]:

$$\phi(r, z) = \frac{Uaj}{2\pi k} \int_{-\pi}^{\pi} \frac{e^{-jkz} - e^{-jk\sqrt{R^2+z^2}}}{a - re^{j\psi}} d\psi,$$

(20)

Among them, the particle vibration potential energy of the circular parameter array is evenly distributed on the plane with radius a, the amplitude is U, and the introduced variables R and ψ meet $R^2 = (a - re^{-j\psi})(a - re^{j\psi})$.

The sound pressure at the (r, z) point can be expressed as:

$$p(r,z,t) = -j\rho_0 \omega e^{j\psi t}\phi(r,z),\tag{21}$$

The formula rewrites the original quadratic integral into one integral, which greatly reduces the computational complexity. By calculating each point in (r, z), the sound field distribution of parametric array can be easily obtained.

3. Noise Control Using Adjustable PAL

In this section, we introduce the hardware and experiments employed, and use an adjustable PAL to adjust adaptively for different distances. Based on this, we compare the noise reduction distance, acoustic feedback and noise reduction performance between the adjustable PAL and the traditional VCL. In addition, we also discuss the distribution of quiet areas.

3.1. Sound Field Distribution of PAL

Parametric array loudspeakers produce high directivity audible sound by using the nonlinear propagation effect of ultrasound wave in the air [33]. The directivity of audible sound mainly depends on the directivity of ultrasound, and they are closely related. Formula 22 is the directivity formula of the ultrasound transducer. α is the angle between the projection of the directivity vector on the plane XOY of the transducer and the X axis, and θ is the angle between the directivity vector and the Z axis of the transducer, J_1 is the first-order Bessel functions, $k = 2\pi/\lambda$, $\lambda = c/f$, k is the wave number, λ is the wavelength, (x_i, y_i) is the position of the transducer, a is the radius of the transducer, and n is the number of the transducer. The units of the α, θ angle are rad; the unit of directivity is non-dimensional. The directivity of the transducer array is shown in Figure 3.

$$D(\alpha,\theta) = \left| \frac{2J_1(ka\sin\theta)}{ka\sin\theta} \right| \left| \sum_{i=1}^{n} e^{jk|x_i\sin\theta\sin\alpha + y_i\sin\theta\cos\alpha|} \right|,\tag{22}$$

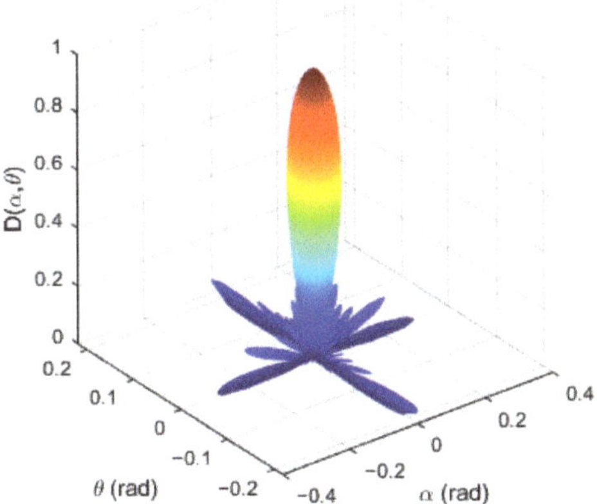

Figure 3. Transducer array directivity. The X-axis is angle α (rad), the Y-axis is angle θ (rad), and the Z-axis is directivity $D(\alpha,\theta)$.

The sound wave propagation of the traditional voice coil loudspeaker is omni-directional, and its direction is 360°. Assuming the emission direction of VCL as the axis, its semi directivity is 180°. The traditional voice coil loudspeakers do not easily produce high directivity audible sound, while the ultrasound transducer has good directivity, but the directivity of a single transducer is not high. The transducer array composed of multiple transducers has stronger directivity, and its side lobe is also effectively suppressed.

The sound field generated by a PAL in air contains ultrasound signals, self-demodulating audible signals. In order to compare the sound field distribution and propagation distance of parametric arrays with different diameters, the composite sound field simulation was carried out according to the parametric array theory. This was performed under standard atmospheric pressure, at a temperature of 20 °C, a relative humidity of 30% RH, a carrier frequency of 40 KHz, and with the diameter of the parameter array at 0.17 m, 0.14 m and 0.11 m. The composite sound field distribution was shown in Figure 4; the unit of amplitude is dB. It can be seen from the figure that sound waves propagated approximately in bundles in the air, and the energy was mainly concentrated in the axial direction. The larger the diameter of the parametric array, the farther the propagation distance. It can be seen from the figure that the parametric array loudspeaker used in the control system had a maximum propagation distance of up to 10 m. The control system was used to eliminate the noise of close range (<10 m) targets. In addition, the smaller the diameter of the parametric array, the smaller the sound field coverage, and the smaller the interference to other adjacent sound fields.

3.2. Adaptive Adjustment of Noise Reduction Target Distance

The circular parametric array loudspeaker used in this paper is shown in Figure 5. The ultrasound frequency is 40 KHz, and the diameter of the parametric array loudspeaker is about 170 mm. The circular array consists of six layers; each layer can be freely controlled on and off. Due to the accumulation effect of parametric array ultrasound beam self-demodulation, the attenuation of audible sound in the propagation direction of the PAL is slower than the traditional VCL, and the propagation distance is longer. The acoustic propagation distance of the PAL is directly proportional to the amplitude of sound pressure and the size of parametric array. Therefore, this paper divided the PAL into three modes (M1 for 1–6 layers, M2 for 1–5 layers, M3 for 1–4 layers), as shown in Figure 5b. The number of ultrasound transducers corresponding to each mode of PAL is different. The larger the size, the farther the sound wave propagation distance, and the greater the corresponding power. We carried out experiments on them in turn.

As the size of PAL increases, the corresponding propagation distance increases successively. The three modes of parametric array loudspeakers play white noise, and the axial sound pressure distribution is measured with a sound level meter, with an axial interval of 0.2 m. Its axial sound pressure distribution is shown in Figure 6a. Before the operation of the ANC system, the secondary source PAL sends ultrasound signal, and the error microphone synchronously detects the arrival time of the pulse, so as to calculate the distance between the error microphone and the secondary source, that is, the noise reduction target distance. According to the distance of the target, the size of the PAL can be freely controlled and the control distance can be roughly adjusted. If the target is far away, more layers are opened. If it is near, the outer layer is closed and only the inner layer is opened. When the same numbers of layers are opened, the fine adjustment of control distance can be realized by adjusting the power. Through the above strategy of coarse adjustment and fine adjustment, the noise reduction control distance of different targets can be adjusted adaptively. For short-range targets, due to the small sound field distribution range of the PAL with a smaller diameter, the use of a PAL with a smaller diameter can avoid the interference to the surrounding adjacent sound field.

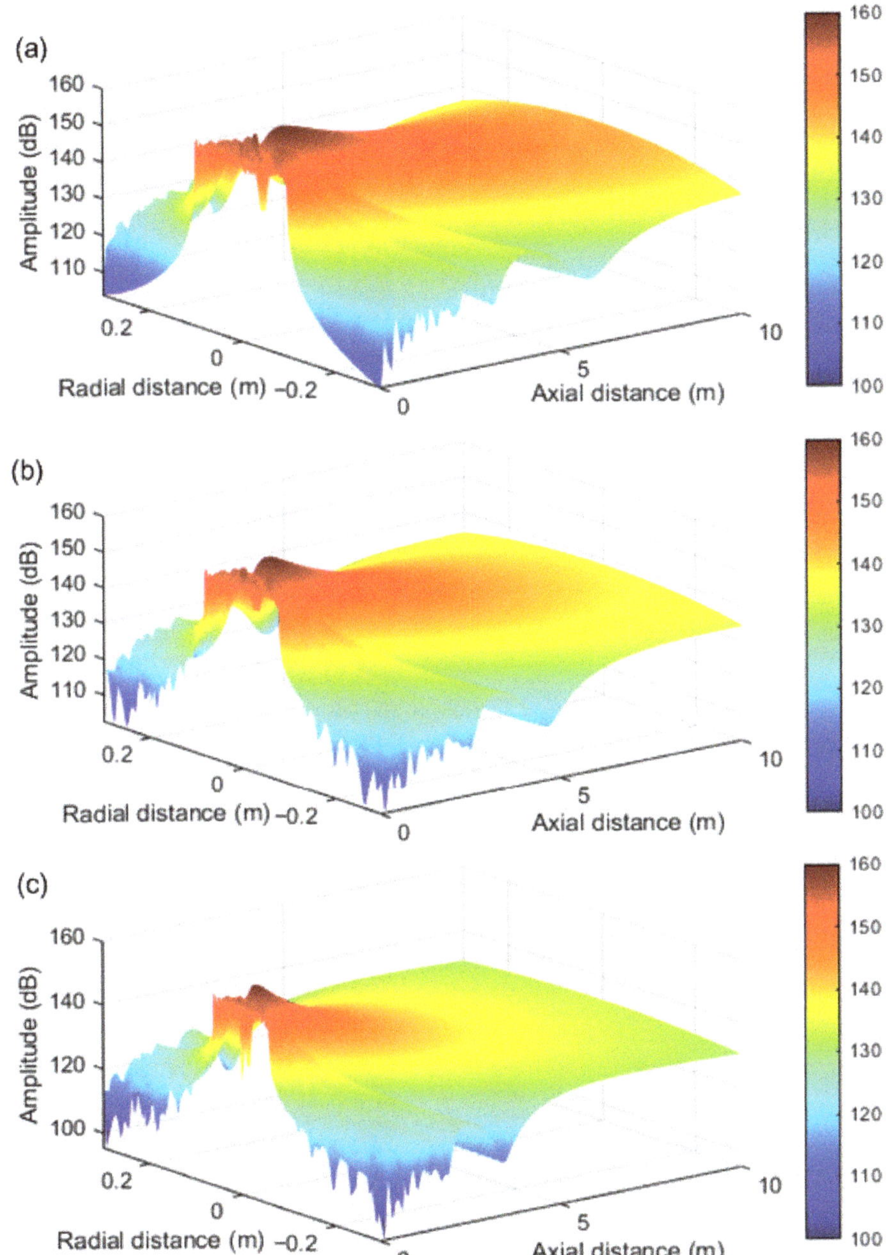

Figure 4. Sound field distribution of parametric array loudspeaker with different diameters. (**a**) 0.17 m in diameter; (**b**) 0.14 m in diameter; (**c**) 0.11 m in diameter.

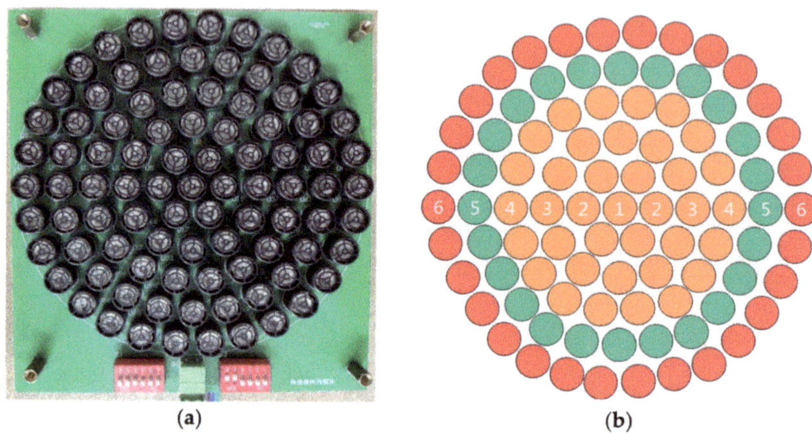

Figure 5. Adjustable PAL: (**a**) physical diagram; (**b**) schematic diagram.

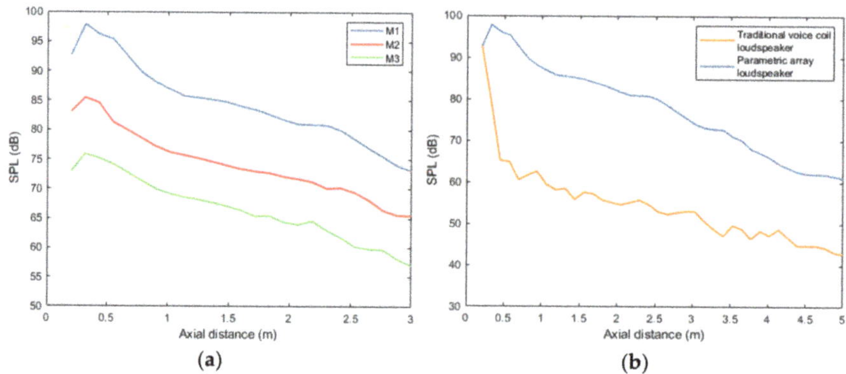

Figure 6. Axial sound pressure variation of traditional VCL and PAL: (**a**) axial sound pressure distribution of PAL in three modes; (**b**) axial sound pressure distribution of traditional VCL and PAL.

In order to further compare the attenuation of sound pressure with distance, the PAL and traditional VCL played white noise. Their initial sound pressure was the same and their axial sound pressure distribution was measured, with an axial interval of 0.2 m. As shown in Figure 6b, with the increase in axial distance, the sound pressure attenuation of the traditional VCL was much larger than the PAL, and the average sound pressure difference between them was about 23.1 dB. Compared with the traditional VCL, the sound pressure of the PAL decreased slowly with the distance, and the propagation distance was longer. Using this characteristic, the PAL can achieve long-distance noise control.

3.3. Acoustic Feedback of Traditional VCL and PAL

Traditional VCLs and PALs were used as secondary source in turn to emit a white noise signal. The reference microphone and error microphone were placed as shown in Figure 7b. The distance between the secondary source and the error microphone was 3 m. All placement positions remained unchanged, and their corresponding signals were measured synchronously, corresponding to the feedback path H(z) and secondary path S(z) in Figure 7a. The DSP control platform with TMS320C6748 of TI Company as the core was adopted, and the signal sampling rate was 8 KHz. The feedback path and secondary path were identified by LMS algorithm. The number of taps of LMS filter was 128 and the update step was 0.001.

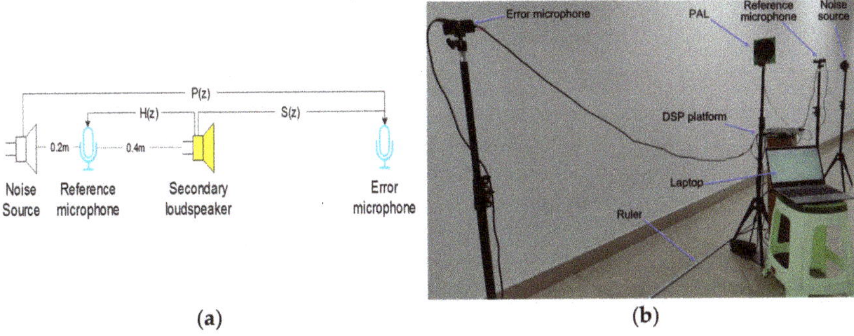

(**a**)   (**b**)

Figure 7. Experiment with laying scenes: (**a**) schematic diagram; (**b**) specific experimental scenarios.

The amplitude frequency diagram of feedback path H(z) and secondary path S(z), corresponding to the traditional VCL and PAL, was shown in Figure 8. It can be seen from the figure that when the PAL was used as the secondary source, the amplitude of the feedback path was obviously smaller than the secondary path. The feedback path of the PAL was much smaller than that of the traditional VCL. As a secondary source, PALs can significantly reduce the sound feedback and make the system more stable.

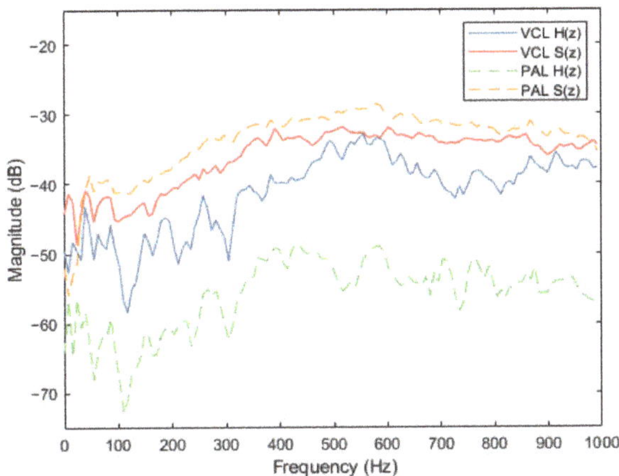

Figure 8. Amplitude frequency response of secondary path S(z) and feedback path H(z) of traditional VCL and PAL.

3.4. Noise Reduction Performance of Traditional VCL and PAL

The experimental scenes are arranged according to Figure 7. Online secondary path modeling is carried out according to the noise reduction target. The DSP control platform with TMS320C6748 of TI Company as the core is adopted, and the signal sampling rate is 8 KHz. The step factor of ANC filter and secondary path filter is 0.001, and the tap length of the filter is 128. The noise source frequency is selected as 400 Hz, 600 Hz and 1000 Hz, respectively, and the narrowband multi-frequency noises are selected as 400 Hz + 600 Hz + 1000 Hz. In this paper, noise control is carried out in turn, and the distribution of noise reduction areas at a single frequency of 600 Hz was discussed.

The noise reduction distribution of two kinds of secondary sources when the noise source is at the single frequencies of 400 Hz and 600 Hz was shown in Figure 9. When the single frequency was 400 Hz, the noise reduction of the PAL and traditional VCL was 15.1 dB and 14.4 dB. When the single frequency was 600 Hz, the noise reduction was 10.8 dB and 12.1 dB, respectively. When the noise source was at the single frequency of

1000 Hz and at multi-frequency, the noise reduction effect was shown in Figure 10. When the single frequency was 1000 Hz, the noise reduction was 11.4 dB and 12.0 dB.. The average noise reduction was 9.7 dB and 10.2 dB at multi-frequency. At single frequency and multi-frequency, the PAL and traditional VCL had the same active noise control effect. It could be seen from the two figures that the fundamental noise amplitude of PAL was higher than the traditional VCL. This was due to the accumulation effect of the PAL beam self-demodulation, which led to the reduction of fundamental frequency noise reduction. In addition, due to the influence of nonlinearity in the air, multiple harmonics will be generated in the signal propagation process, which will interfere with the effective signal and reduce the signal quality. When the noise is a narrow-band signal of 600 Hz–1200 Hz, there is no significant difference between the noise reduction effect of the parametric array loudspeaker and the traditional voice coil loudspeaker. The average noise reduction of the two is 11.5 dB and 12.3 dB, respectively. Therefore, the parametric array loudspeaker and the traditional voice coil loudspeaker have the same noise reduction effect as the secondary source in the active noise control system.

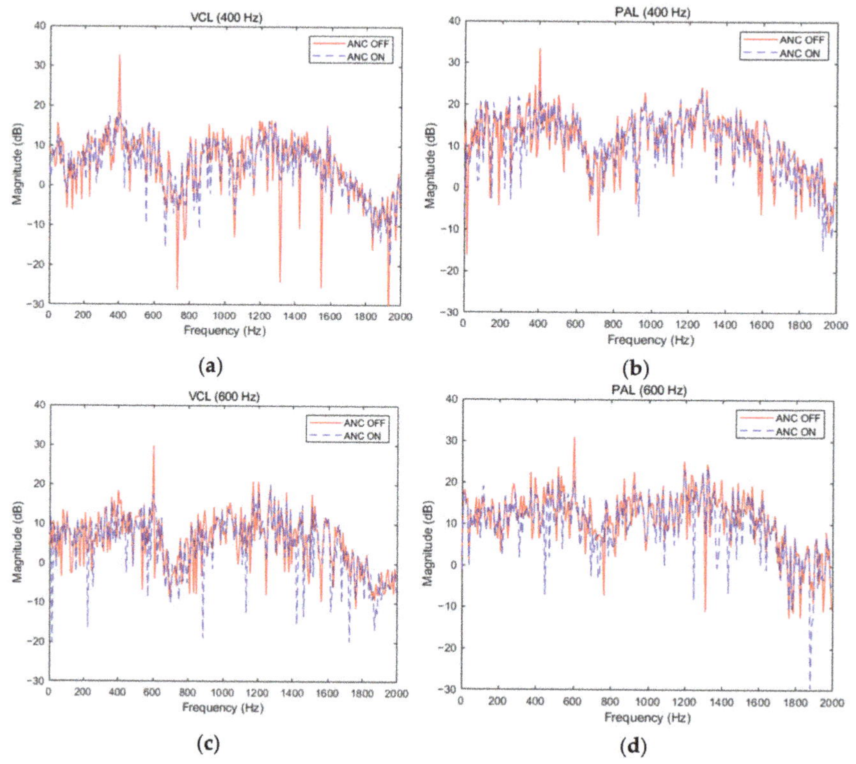

Figure 9. Noise reduction effect of traditional VCL and PAL at 400 Hz and 600 Hz. (**a**) VCL at 400 Hz; (**b**) PAL at 400 Hz; (**c**) VCL at 600 Hz; (**d**) PAL at 600 Hz.

3.5. The Distribution of Noise Reduction Area

In order to compare the noise reduction areas distribution of traditional VCL and PAL, the noise reduction areas of them were measured. The noise reduction experiments with the above noise source at 600 Hz were taken as the object. The noise reduction distribution around the error microphone was measured. As shown in Figure 11, in a 0.8 m × 1.6 m rectangular area perpendicular to the ground, the sound pressure was measured at equal intervals with a particle size of 10 cm. A total of 306 sound pressure values were measured before and after the noise reduction, so as to intuitively reflect the change trend of the rectangular noise area before and after noise reduction. The "+" in Figure 11 was the position of the error microphone, and the noise source, secondary source and error microphone were in a straight line, as shown in Figure 7.

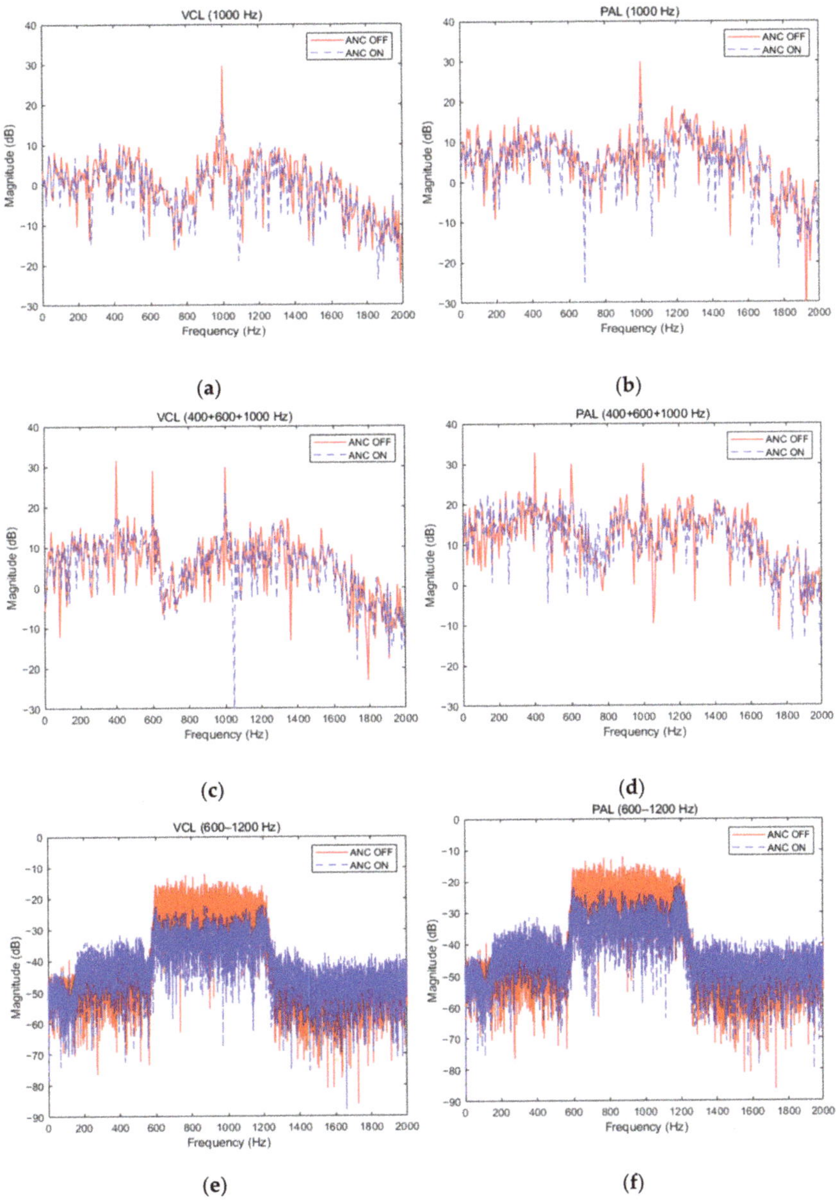

Figure 10. Noise reduction effect of traditional VCL and PAL at single frequency 1000 Hz, multi-frequency (400 Hz + 600 Hz + 1000 Hz) and narrow-band frequency (600 Hz–1200 Hz). (**a**) VCL at 1000 Hz; (**b**) PAL at 1000 Hz; (**c**) VCL at multi-frequency (400 Hz + 600 Hz + 1000 Hz); (**d**) PAL at multi-frequency (400 Hz + 600 Hz + 1000 Hz); (**e**) VCL at narrow-band frequency (600 Hz–1200 Hz); (**f**) PAL at narrow-band frequency (600 Hz–1200 Hz).

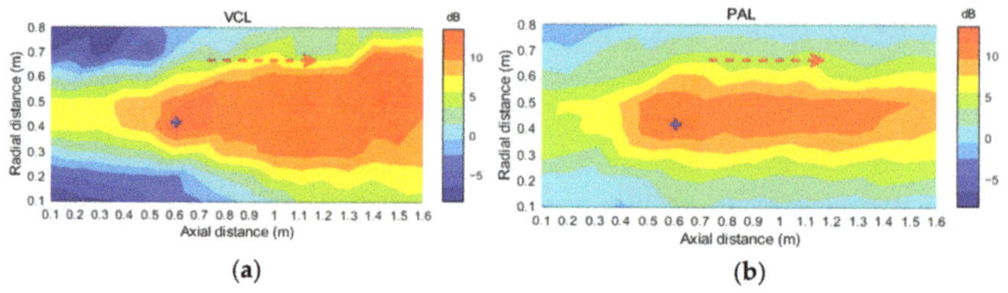

Figure 11. Noise reduction distribution of PAL and traditional VCL. (**a**) VCL; (**b**) PAL.

As shown in Figure 11, taking the error microphone as the starting point, a certain range of quiet area was established along the sound propagation direction (as indicated by the dashed red arrow). The noise reduction area of the PAL as secondary source was smaller than that of the traditional VCL. The boundary distribution of noise reduction areas of PAL was more controllable and approximately rectangular, while the traditional VCL was approximately fan-shaped. The PAL had obvious advantages in the application of noise reduction in a specific range. In the adjacent sound pressure distribution, the increment of the traditional VCL to other areas was 9 dB, while the increment of the PAL to adjacent areas was less than 2 dB. Due to noise control, the noise in other adjacent areas increases, so that people are exposed to a high noise environment, which will bring risks to people's hearing [34]. Compared with the omni-directionality of the traditional voice coil loudspeaker, the parametric array loudspeaker has high directivity. When it is used as the secondary source, the interference of the reverse sound wave emitted by it to other adjacent areas can be ignored, and it does not cause the increase of sound pressure level in other areas. The sound field of the PAL is mainly concentrated in the axial area, and the area of its quiet area is smaller than that of the VCL, but the noise reduction area is mainly concentrated in the axial area, which is more controllable. Therefore, the PAL had the advantages of controllable noise reduction areas and small interference to adjacent areas. However, the current technique also has limitations, such as the control system producing harmonic distortion and fundamental frequency noise. It can be further solved by optimizing carrier signal modulation methods, such as improving signal broadband and spectrum utilization. The control system needs to place a real microphone at the target point to collect residual noise. In the future, the virtual microphone technology can be used to replace the real microphone, which has simplified the control system.

4. Conclusions

In this paper, the adjustable PAL is used as the secondary source emitting reverse noise waves to eliminate environmental noise, and the size and power of the PAL are adjusted adaptively according to the different noise target, so as to realize the active noise control in the target area. The secondary path modeling with variable power auxiliary noise can realize adaptive noise reduction for different long-range targets and make the system converge faster and have a lower steady-state offset. For indoor environments, the sound pressure attenuation of the traditional VCL was much greater than the PAL's, which limited noise control of the traditional VCL to the short-range target only. The experimental results showed that by adjusting the power and radius of the PAL, the noise control of different distance targets could be realized. The PAL had the same noise reduction performance as the traditional VCL, and the maximum noise reduction was 15.1 dB. In the noise reduction areas distribution, although the PAL was smaller than the traditional VCL, the noise interference to other adjacent areas could be ignored, and the noise reduction areas were more controllable. These results have some practical guiding significance for environmental noise cancellation at long distance and specific range in indoor environments.

Author Contributions: Conceptualization, methodology, Y.L. and W.Z.; software, validation, investigation, writing—original draft preparation, Y.L.; resources, writing—review and editing, W.Z. All authors have read and agreed to the published version of the manuscript.

Funding: This research was funded by (National Natural science Foundation of China) grant number (61573073).

Institutional Review Board Statement: Not applicable.

Informed Consent Statement: Not applicable.

Data Availability Statement: The datasets used and/or analyzed during the current study are available from the corresponding author on reasonable request.

Acknowledgments: We would like to thank Zhengqiang Luo for his profound discussion on sound field theory, as well as for his help in setting up the experiment.

Conflicts of Interest: The authors declare no conflict of interest.

References

1. Wouters, N.L.; Kaanen, C.I.; den Ouden, P.J.; Schilthuis, H.; Bohringer, S.; Sorgdrager, B.; Ajayi, R.; de Laat, J. Noise Exposure and Hearing Loss among Brewery Workers in Lagos, Nigeria. *Int. J. Environ. Res. Public Health* **2020**, *17*, 2880. [CrossRef] [PubMed]
2. Trieu, B.L.; Nguyen, T.L.; Hiraguri, Y.; Morinaga, M.; Morihara, T. How Does a Community Respond to Changes in Aircraft Noise? A Comparison of Two Surveys Conducted 11 Years Apart in Ho Chi Minh City. *Int. J. Environ. Res. Public Health* **2021**, *18*, 4307. [CrossRef] [PubMed]
3. Lee, H.M.; Hua, Y.T.; Wang, Z.M.; Lim, K.M.; Lee, H.P. A review of the application of active noise control technologies on windows: Challenges and limitations. *Appl. Acoust.* **2021**, *174*, 7. [CrossRef]
4. Zhang, J.H.; Abhayapala, T.D.; Zhang, W.; Samarasinghe, P.N.; Jiang, S.D. Active Noise Control Over Space: A Wave Domain Approach. *IEEE-ACM Trans. Audio Speech Lang.* **2018**, *26*, 774–786. [CrossRef]
5. Chu, Y.J.; Mak, C.M.; Zhao, Y.; Chan, S.C.; Wu, M. Performance analysis of a diffusion control method for ANC systems and the network design. *J. Sound Vib.* **2020**, *475*, 17. [CrossRef]
6. Hirose, S.; Kajikawa, Y.; IEEE. Effectiveness of Headrest ANC System with Virtual Sensing Technique for Factory Noise. In Proceedings of the 2017 Asia-Pacific Signal and Information Processing Association Annual Summit and Conference, Aloft Kuala Lumpur Sentral, Kuala Lumpur, Malaysia, 12–15 December 2017; IEEE: New York, NY, USA, 2017; pp. 464–468.
7. Yang, F.R.; Guo, J.F.; Yang, J. Stochastic Analysis of the Filtered-x LMS Algorithm for Active Noise Control. *IEEE-ACM Trans. Audio Speech Lang.* **2020**, *28*, 2252–2266. [CrossRef]
8. Burgess, J.C. Active adaptive sound control in a duct—A computer-simulation. *J. Acoust. Soc. Am.* **1981**, *70*, 715–726. [CrossRef]
9. Eriksson, L.J.; Allie, M.C. Use of random noise for online transducer modeling in an adaptive active attenuation system. *J. Acoust. Soc. Am.* **1989**, *85*, 797–802. [CrossRef]
10. Akhtar, M.T.; Abe, M.; Kawamata, M. A new variable step size LMS algorithm-based method for improved online secondary path modeling in active noise control systems. *IEEE Trans. Audio Speech Lang. Process.* **2006**, *14*, 720–726. [CrossRef]
11. Carini, A.; Malatini, S. Optimal variable step-size NLMS algorithms with auxiliary noise power scheduling for feedforward active noise control. *Ieee Trans. Audio Speech Lang. Process.* **2008**, *16*, 1383–1395. [CrossRef]
12. Ahmed, S.; Akhtar, M.T.; Zhang, X. Robust Auxiliary-Noise-Power Scheduling in Active Noise Control Systems With Online Secondary Path Modeling. *IEEE Trans. Audio Speech Lang. Process.* **2013**, *21*, 13. [CrossRef]
13. Aslam, M.S.; Raja, M.A.Z. A new adaptive strategy to improve online secondary path modeling in active noise control systems using fractional signal processing approach. *Signal Process.* **2015**, *107*, 433–443. [CrossRef]
14. Elliott, S.J.; Boucher, C.C.; Nelson, P.A. The behavior of a multiple channel active control-system. *IEEE Trans. Signal Process.* **1992**, *40*, 1041–1052. [CrossRef]
15. Aslan, F.; Paurobally, R. Modelling and simulation of active noise control in a small room. *J. Vib. Control* **2018**, *24*, 607–618. [CrossRef]
16. Elliott, S.J.; Jones, M. An active headrest for personal audio. *J. Acoust. Soc. Am.* **2006**, *119*, 2702–2709. [CrossRef]
17. Liu, N.N.; Sun, Y.D.; Wang, Y.S.; Sun, P.; Li, W.W.; Guo, H. Mechanism of interior noise generation in high-speed vehicle based on anti-noise operational transfer path analysis. *Proc. Inst. Mech. Eng. Part D -J. Automob. Eng.* **2021**, *235*, 273–287. [CrossRef]
18. Aslam, M.S.; Shi, P.; Lim, C.C. Self-adapting variable step size strategies for active noise control systems with acoustic feedback. *Automatica* **2021**, *123*, 9. [CrossRef]
19. Ahmed, S.; Akhtar, M.T. Gain Scheduling of Auxiliary Noise and Variable Step-Size for Online Acoustic Feedback Cancellation in Narrow-Band Active Noise Control Systems. *IEEE-ACM Trans. Audio Speech Lang.* **2017**, *25*, 333–343. [CrossRef]
20. Komatsuzaki, T.; Iwata, Y. Active Noise Control Using High-Directional Parametric Loudspeaker. *J. Environ. Eng.* **2011**, *6*, 140–149. [CrossRef]
21. Shi, C.; Kajikawa, Y.; Gan, W.S. An overview of directivity control methods of the parametric array loudspeaker. *APSIPA Trans Sig. Inf. Process.* **2014**, *3*, 1–12. [CrossRef]

22. Tanaka, K.; Shi, C.; Kajikawa, Y. Multi-channel Active Noise Control Using Parametric Array Loudspeakers. In Proceedings of the 2014 Asia-Pacific Signal and Information Processing Association Annual Summit and Conference, Angkor, Cambodia, 9–12 December 2014; IEEE: New York, NY, USA, 2014; pp. 1–6.
23. Brooks, L.A.; Zander, A.C.; Hansen, C.H. Investigation into the Feasibility of Using a Parametric Array Control Source in an Active Noise Control System. In Proceedings of the ACOUSTICS 2005, Busselton, WA, Australia, 9–11 November 2005; pp. 39–45.
24. Kidner, M.; Petersen, C.; Zander, A.C.; Hansen, C.H. Feasibility Study of Localised Active Noise Control Using an Audio Spotlight and Virtual Sensors. In Proceedings of the ACOUSTICS 2006, Christchurch, New Zealand, 20–22 November 2006; pp. 55–61.
25. Tanaka, K.; Shi, C.; Kajikawa, Y. Study on Active Noise Control System Using Parametric Array Loudspeakers. In Proceedings of the 7th Forum Acusticum, Krakow, Poland, 7–12 September 2014.
26. Ganguly, A.; Vemuri, H.K.; Panahi, I. Real-Time Remote Cancellation of Multi-Tones in an Extended Acoustic Cavity Using Directional Ultrasonic Loudspeaker. In Proceedings of the Iecon 2014—40th Annual Conference of the IEEE Industrial Electronics Society, Dallas, TX, USA, 29 October–1 November 2014; IEEE: New York, NY, USA, 2014; pp. 2445–2450. [CrossRef]
27. Tanaka, K.; Shi, C.; Kajikawa, Y. Binaural active noise control using parametric array loudspeakers. *Appl. Acoust.* **2017**, *116*, 170–176. [CrossRef]
28. Akhtar, M.T.; Abe, M.; Kawamata, M. Noise power scheduling in active noise control systems with online secondary path modeling. *IEICE Electron. Express* **2007**, *4*, 66–71. [CrossRef]
29. Zabolotskaya, E.A.; Khokhlov, R.V. Quasi-Plane waves in nonlinear acoustics of cofined beams. *Soviet Physics Acoust. -Ussr* **1969**, *15*, 35–40.
30. Kuznetsov, V.P. Equations of nonlinear acoustics. *Sov. Phys. Acoust. -Ussr* **1971**, *16*, 467–470.
31. Hamilton, M.F.; Blackstock, D.T. *Nonlinear Acoustics: Theory and Applications*; Academic Press: Washington, DC, USA, 1997.
32. Liauh, C.T.; Lin, W.L. Fast numerical scheme of computing acoustic pressure fields for planar circular ultrasound transducers. *J. Acoust. Soc. Am.* **1999**, *105*, 2243–2247. [CrossRef]
33. Westervelt, P.J. Parametric acoustic array. *J. Acoust. Soc. Am.* **1963**, *35*, 535–537. [CrossRef]
34. Rahim, K.A.A.; Jewaratnam, J.; Hassan, C.R.C.; Hamid, M.D. Effectiveness of a Novel Index System in Preventing Early Hearing Loss among Furniture Industry Skills Training Students in Malaysia. *Int. J. Environ. Res. Public Health* **2020**, *17*, 16. [CrossRef]

International Journal of
Environmental Research and Public Health

Article

Helicopter Inside Cabin Acoustic Evaluation: A Case Study—IAR PUMA 330

Marius Deaconu [1], Grigore Cican [2,*], Adina-Cristina Toma [1] and Luminița Ioana Drăgășanu [1]

1   National Research and Development Institute for Gas Turbines COMOTI, 220D Iuliu Maniu, 061126 Bucharest, Romania; marius.deaconu@comoti.ro (M.D.); adina.toma@comoti.ro (A.-C.T.); luminita.dragasanu@comoti.ro (L.I.D.)
2   Department of Aerospace Sciences, Faculty of Aerospace Engineering, Polytechnic University of Bucharest, 1-7 Polizu Street, 1, 011061 Bucharest, Romania
*   Correspondence: grigore.cican@upb.ro

Citation: Deaconu, M.; Cican, G.; Toma, A.-C.; Drăgășanu, L.I. Helicopter Inside Cabin Acoustic Evaluation: A Case Study—IAR PUMA 330. *Int. J. Environ. Res. Public Health* **2021**, *18*, 9716. https://doi.org/10.3390/ijerph18189716

Academic Editors: Peter Lercher, Gaetano Licitra and Luca Fredianelli

Received: 16 July 2021
Accepted: 8 September 2021
Published: 15 September 2021

Publisher's Note: MDPI stays neutral with regard to jurisdictional claims in published maps and institutional affiliations.

Abstract: This paper presents an inside-cabin acoustic evaluation of the IAR PUMA 330 helicopter, manufactured by IAR S.A. Brasov. In this study, based on the acoustic assessment inside the helicopter, areas with high noise levels are identified. In this regard, several tests were carried out in accordance with the ISO 5129 standard. In the first stage of the assessment, a measurement campaign was performed to identify the acoustic leaks from the outside noise sources propagating inside the cabin (in the door area) and the acoustic attenuation of the helicopter structure. These tests were performed on the factory runway, with the helicopter in parked position (ground tests). During the ground tests, the helicopter engines were turned off. The tests consisted of placing two loudspeakers directed towards the helicopter door and generating pink noise. Inside the helicopter, the entire door frame opening was scanned with an intensity probe to identify acoustic leaks areas. The second assessment stage was to determine the areas of the cabin with the highest levels of noise. Within the measurement campaign, 16 microphones were placed inside the cabin, at the level of the passengers' heads, arranged in seven zones. The tests were carried out with the helicopter engines started, staying at fixed point above the ground (hovering), and then a flight test, in which all the maneuvers necessary for the use of the helicopter were performed (in-flight tests). Based on the measurement results, it was possible to highlight the noise spectral components in each of the seven areas. The noise assessment revealed high noise levels inside the cabin, having as main noise sources the transmission gear and the door area, leading to the need for reducing the noise exposure for passengers and crew, thus the need to reduce noise levels inside the helicopter.

Keywords: helicopter cabin; noise levels; noise reduction; acoustic evaluation; IAR Puma 330

1. Introduction

Nowadays, noise pollution represents one of the main problems for aviation development [1] due to the need to heavily reduce the noise exposure of the areas adjacent to airports or heliports. It is known that the noise exposure to the crew and especially the passengers and the people around landing and take-off areas represents a great issue [2].

In terms of helicopter noise sources, the main sources are: rotor, anti-torque, engines, gear box, depending on flight condition, transmission gear, etc. [3] and are illustrated in Figure 1.

In a short review of the noise sources generated by the helicopter and connected to the present study, it is worth mentioning the thickness noise which is caused by the blade periodically displacing air during each revolution and is dependent only on the shape and motion of the blade. Generally, the thickness noise propagates in the plane of the rotor as well as the high-speed impulsive noise. In addition, the loading noise is another type of noise source which influences the inside-cabin noise. The loading noise is directed below the rotor and is caused by the acceleration of the force distribution on the air around the rotor blade passing through it. Another influence is the blade vortex interaction which is directed down and rearward and it occurs when a rotor blade passes within close proximity of the shed tip vortices from a previous blade [4,5].

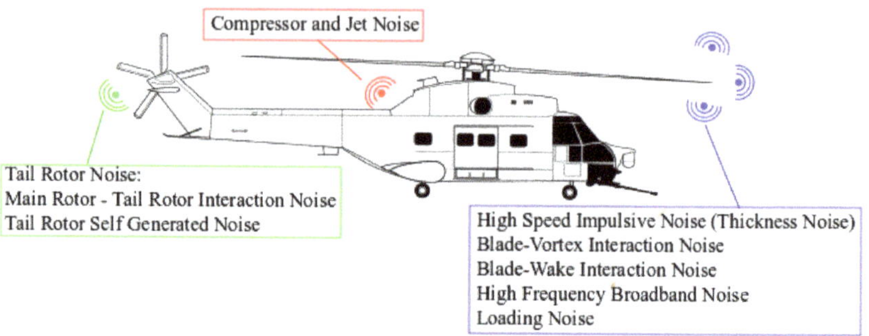

Figure 1. Helicopter noise sources, generation locations and their classification.

From the frequency point of view, the most annoying noise for the human ear is the tail rotor noise due to its higher frequency which coincides with the band to which the human ear is most sensitive [6–8].

The inside-cabin noise levels depend on the flight conditions, maneuvers and the observer position inside the helicopter. All these combined influences are detailed by the joint work of Snecma, Airbus Helicopters, Sikorsky Aircraft, Bell Helicopter, Agusta Westland, Turbomeca, Marenco Swisshelicopter and the Research Centers: NASA, DLR, ONERA, JAXA in [9]. This comprehensive study is highlighting the fact that implementing a sophisticated noise reduction technology addressing one noise source may reduce the noise level in one flight condition; there may be, however, no change or in some cases increases in the noise levels in other flight conditions [9].

Helicopters with a good acoustic level are considered to be those which have 70 dB(A) in the interior, while those in which the noise exposure exceeds 85 dB(A) equivalent sound level are considered to be a potential risk [10].

The purpose of this paper is essentially to study the acoustic field inside the cabin, to determine the inside noise level, whether soundproofing structures are needed, and to determine to what frequency domain the structures must be designed.

It is very important to have the complete noise description inside the helicopter cabin, in order to know the exposure level and to find ways to improve the noise conditions.

A similar paper that approaches the acoustic evaluation of helicopters and particularly focuses on acoustical comfort improvement in helicopter cabins is presented in [11]. There are available, and already performed, different methods to characterize the inside-cabin noise, and also methods used to localize noise sources inside helicopter cabin from in-flight tests [12,13]. Several studies have been conducted regarding helicopter noise such as: acoustic performance measurements during flight for a Bell 206B helicopter [14], exterior noise level induced by two different rotor blades for two helicopters, R44 Clipper and Bell 206B [15]; and another study presents the results of measurements of noise generated by helicopter SA341H "Gazelle" at full throttle in the landing and take-off phase [16]. Here, the level of noise generated by helicopter type SA341H "Gazelle" was compared to the levels of noise generated by the Robinson R44 Clipper and Bell 206B helicopters. Nelson [17] presents acoustic measurements and data collected inside a U.S. Army Sikorsky UH-60 helicopter to analyze the inherent noise present during routine aeromedical transport.

Addressing a similar goal to our research study is the work performed by Eurocopter, the difference between that work and our research being the fact that they used a modelling methodology coupled with Nearfield Acoustical Holography measurements and geometrical acoustics [18].

The acoustic evaluation of the IAR PUMA 330 helicopter was realized following the ISO 5129 standard. For the acoustic measurement, 16 microphones were placed, according to standard, inside the cabin, at the passenger head level and the noise levels were measured during different maneuvers performed in flight. The noise level variations for each microphone were correlated with the maneuvers during the flight. Another stage consisted of performing tests with the helicopter on the ground with the engine turned off, the noise source being two loudspeakers placed near the helicopter with the purpose of identifying any acoustic leaks from door frames. Finally, conclusions are drawn regarding the level of noise and the inside characteristics of the cabin.

2. IAR PUMA 330 Helicopter Technical Description

IAR PUMA 330 helicopters, Figure 2, are built, maintained and upgraded by the IAR Company S.A. BRASOV [19]. According to the technical datasheet, the crew of this helicopter is composed of three members. The helicopter has a capacity of 16 passengers and a length of 18.22 m. The helicopter has a height of 5.14 m, having a rotor with a diameter of 15.08 m. The empty weight of such a helicopter is 3615 kg, with the take-off maximum of 7400 kg. The helicopter has two TURMO IV C turboshafts with a free turbine, 1,175,000 W each one. The helicopter reaches a maximum speed of 263 km/h and has a range of action of 550 kilometers without other supplementary tanks. The technical features include a 4800 m service cap and an ascending speed of 9.2 m/s [20]. IAR PUMA 330s are helicopters of the 1970s, built to the military specifications of the time when the rules for cab noise were more permissive.

Figure 2. IAR PUMA 330 Helicopter.

3. Testing Procedures

The method used to determine the noise level inside the helicopter during the flight is given by the international ISO 5129 standard [21], which specifies that the measurements should determine the sound pressure level A-weighted and in 1/3 octave band. In addition, ISO 5129 specifies that the sound pressure levels must be measured at the head level of the passengers, but no passenger must be present during the tests. The measuring positions were chosen to determine the acoustic field in the passengers' position and in the entire helicopter. The microphones were fixed with a metallic extension attached to the helicopter frames and a damping material was used as interface between microphone and metallic support to minimize the vibrations' effects on the acoustic signals. The acoustic measurement was performed over the entire flight time. During the acoustic measurement, the helicopter was in a minimal configuration: without upholstery and chairs. During the flight test, only the crew and the acoustic team were present in the helicopter. The first step of the research was to identify the level of sound pressure inside the cabin of the helicopter using the actual soundproofing structures used by IAR Brasov and the acoustic leaks caused by the imperfections of door tightness. For the noise leak detection, the method specified in SR EN ISO 9614-2/2000 [22] was used; a method that is based on mapping sound intensity over the area of interest.

4. Measurement Campaign

The purpose of the inside acoustic evaluation of the IAR 330 cabin was to identify the locations with high noise levels and the cabin's overall noise level. During the helicopter's functioning, inside the cabin, the main source of noise is represented by the transmission box lid, situated close to the passengers' location. Other sensitive locations that can produce a lot of noise are the door areas due to their sealing system. The noise emitted by the exhaust engines (situated above both doors), combined with the aeroacoustics noise, come through the doors' weather-strips. It must be mentioned that the turboshaft engines are fixed on a rigid metal plate that has no physical contact with the cabin indoors.

For noise mapping and sound field characterization, the measurement campaign consists of two steps: ground tests (acoustic measurements with helicopter engines and auxiliary units off) and in-flight tests, one in hovering mode and one performing all the maneuvers necessary for the use of the helicopter.

4.1. Flight Tests

The inside cabin acoustic field was measured by mounting 16 microphones in different locations of the helicopter as is presented in Figure 3. The microphone grid was composed from 16 diffuse field microphones 40AQ type with preamplifier 26CA, which were mounted on special metallic supports, attached to the helicopter frames according to Figure 3. The acoustic signals were recorded with the multichannel acquisition system Sirius from DeweSoft using a sampling frequency of 50 ks/s (kilosamples per second). The calibration of the measurement channels was performed with the acoustic calibrator 42AB GRAS which generates at the frequency of 1 kHz an amplitude of 10.02 Pa (114 dB (ref. 2.0×10^{-5} Pa)). The helicopter cabin was divided in 7 regions corresponding to the structural frames on which were placed the microphone holders as can be seen in Figure 4.

For the turbo engines influence, areas 1 and 2 were designated, area 3 for transmission gear, area 4 for the combined noise emitted by the transmission gear and exhaust noise, and areas 5, 6 and 7 to identify spectral components of other noise sources.

The flight test consisted in performing the following phases: engines start, hovering, hovering turn, hovering—forward, sidewards, rearward flight, turns, climb, descent.

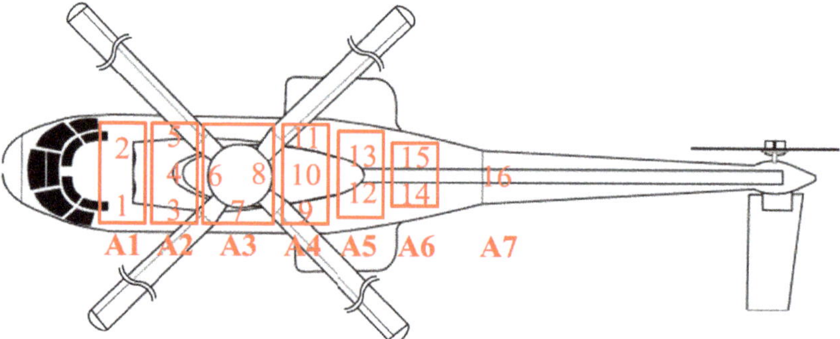

Figure 3. The helicopter cabin divided in 7 areas (A1 to A7) and microphone positions from 1 to 16.

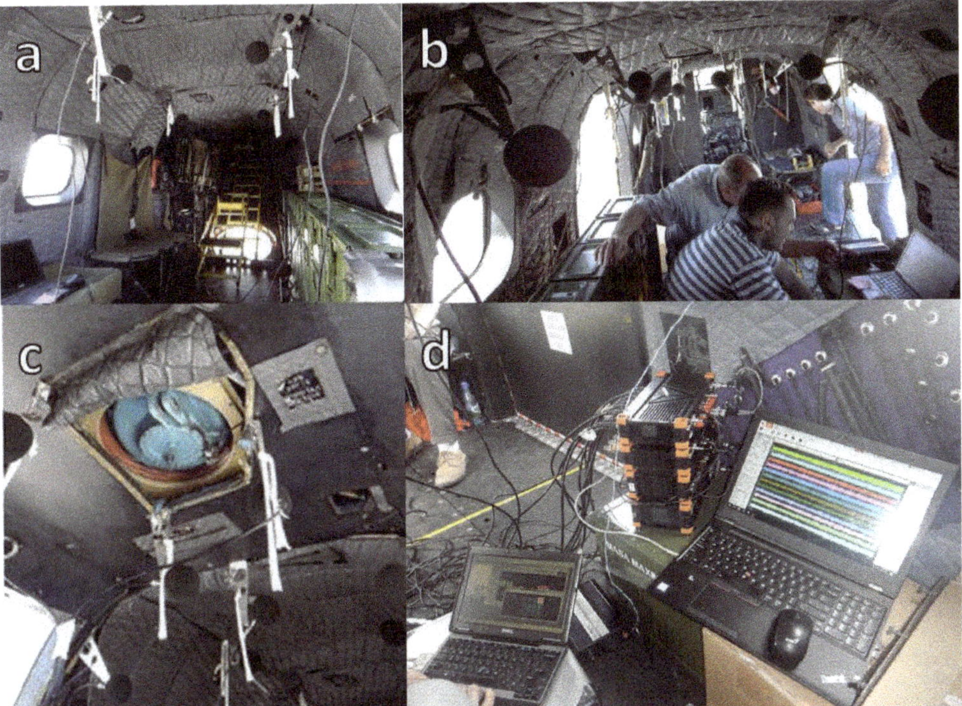

Figure 4. The locations of the microphones in the helicopter and the measurement equipment: (**a**) microphones from A5, A6, A7; (**b**) microphones from A1, A2, A3, A4; (**c**) microphones under transmission gear A3; (**d**) measurement equipment.

4.2. Ground Tests

The purpose of the tests performed on the ground was to identify the acoustic leaks from the doors' weather-strips and to determine the acoustic attenuation of the helicopter side structure (IL—insertion loss) with the current acoustic insulation solution of the IAR. The acoustic tests from the ground were conducted with the helicopter in flight configuration without the engines or other components working. The helicopter was placed on the IAR Brasov runway, with no other acoustic sources near to it during the measurements.

Considering the great distances from the closest nearby buildings, it was considered that during tests the free field condition was reached, so no influencing reflections were considered. At 1.5 m from the helicopter fuselage and 5 m from the loudspeakers, four 40AE microphones were mounted at 1.7 m height from the ground, oriented to loudspeakers. The outside microphones were used to check if the acoustic field generated by the two loudspeakers was diffuse.

Three 40AQ microphones were mounted inside the helicopter, one in the front area of the helicopter, the second one in the middle and the third one at the back. Based on the average sound pressure levels calculated based on the acoustic signals recorded with the outside microphones and the average sound level from the inside microphones, the insertion loss (IL) was calculated.

Before the field measurement campaign, an acoustic evaluation field was performed in the anechoic chamber [23] to determine the number of loudspeakers, their positions and at what distance should be used to obtain a diffuse field. The anechoic room volume was 1200 m^3 with $15 \times 10 \times 8$ m, wall absorption coefficient was 99% in frequency range of 50 Hz up to 20,000 Hz.

For these tests four HK Audio Linear L5 112 F loudspeakers and one power unit LD Systems DP2400X were used. Figure 5 presents one configuration of loudspeaker positioning during the anechoic measurements and the obtained noise spectra in each microphone.

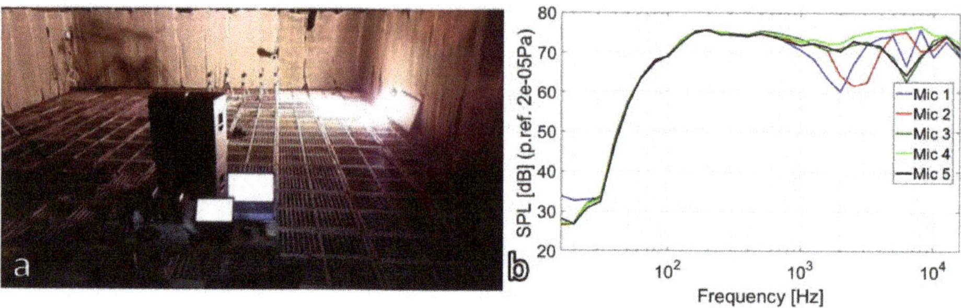

Figure 5. Diffuse field measurement set-up in anechoic chamber and obtained acoustic spectra: (**a**) microphones linear array and loudspeakers position in anechoic chamber; (**b**) averaged noise spectra in each microphone.

Following the tests, it was found that the use of two loudspeakers rotated in different directions by about five degrees can generate a diffuse acoustic field at 5 m.

The spectral analysis presented in Figure 5 highlights that in the frequency domain of 50–1 kHz, the amplitude spectra of the five microphones have a variation of maximum 1.5 dB, while after 1 kHz the obtained differences between microphones reaches almost 10 dB, highlighting that a diffuse field in a specific frequency domain could be obtained using four loudspeakers.

In parallel to these IL tests, acoustic intensity measurements were performed on the helicopter door area, Figure 6.

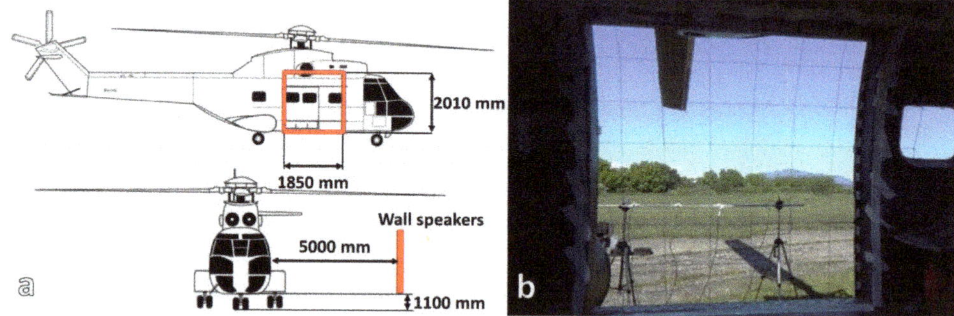

Figure 6. IAR Puma 330 acoustic intensity measurements set-up: (**a**) loudspeakers location, (**b**) outside microphones location and the grid for intensity scanning).

During flight, this area of the door is exposed from the outside of the helicopter to high levels of noise especially that caused by engine exhaust. The door area was divided into 81 surfaces and scanned with the 50AI GRAS intensity probe. The averaging time for each measurement point was 15 s. On the right side of Figure 6 the outside microphones can be observed with which the diffusion of the acoustic field was controlled and monitored.

5. Experimental Results

5.1. Flight Results

Flight results are presented both in the time and frequency domain for the entire duration of the flight. Variation of the overall noise level highlights the influences of the flight conditions. To have an overview of the noise inside the helicopter, an averaged acoustic pressure level for each researched area was calculated according to the standard ISO 3746 [24] using the following equation:

$$L'_{pA} = 10 \, lg \, lg \left[\frac{1}{N} \sum_{i}^{N} 10^{0.1 L'_{pAi}} \right] \; [dB] \tag{1}$$

where L'_{pA} is the average A-weighted sound pressure level, in decibels, with the functioning helicopter; L'_{pAi} represents the A-weighted sound pressure level measured in i position of the microphone in decibels; N is the number of microphone positions.

In Figure 7 the overall averaged noise levels' variation in time is presented.

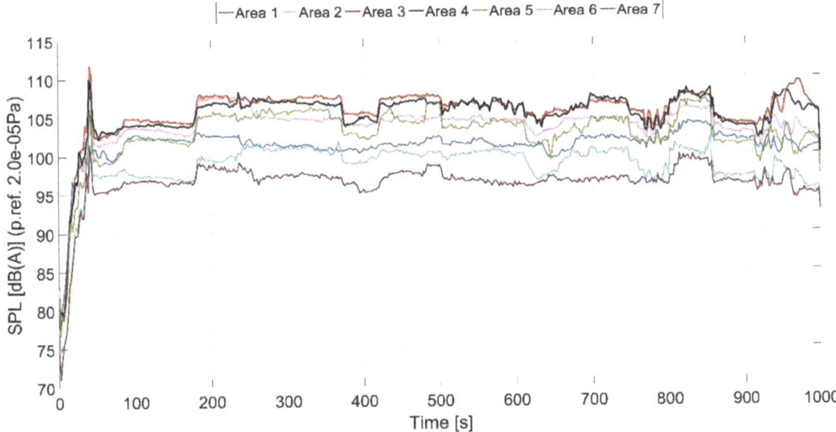

Figure 7. Noise variation in time in each measured area.

From Figure 7 which shows the variation of noise over time, it is observed that the areas with the highest noise level are 3 and 4; the noise level decreases in the helicopter's other areas.

The overall noise level for the entire helicopter was computed based on the averaged noise curves from Figure 7, presented in Figure 8.

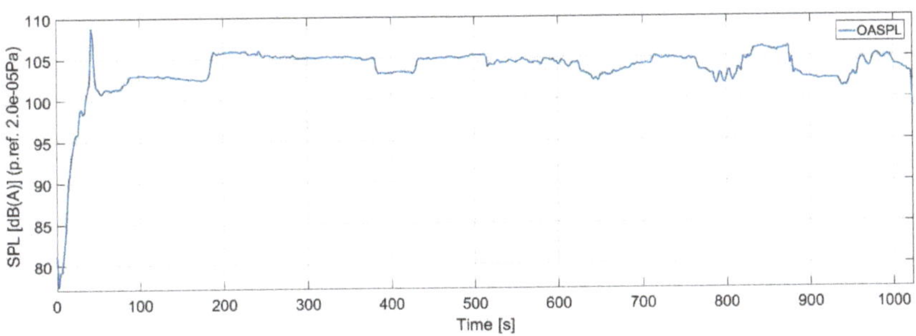

Figure 8. Noise variation for entire helicopter during the flight.

The overall noise level on the entire helicopter calculated according to the above relationship indicates a noise peak reaching 108 dB(A) during start-up, after which the noise level stabilizes to 103 dB(A) with variations depending on the flight conditions as presented in Figure 9. The range of the helicopter flight parameters are presented in Table 1.

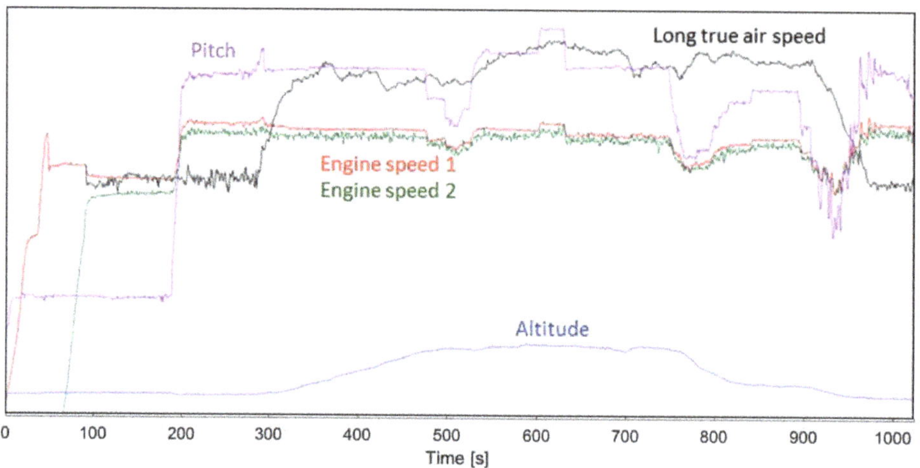

Figure 9. Helicopter flight data over the entire flight.

Table 1. The min & max values during the flight.

Parameter	Min	Max
Altitude [m]	464	1031
Engine 1 speed [%]	0	92.8
Engine 2 speed [%]	0	95.8
Long true air speed [m/s]	0	237
Pitch [deg.]	−2.2	2.5

During the flight the helicopter performed most of the maneuvers as can be seen from the flight data. By comparing the overall noise variation and the flight data, one can observe a correlation between the two sets of data. The starting of the first engine produces a noise peak with the biggest amplitude in time. The noise is related for the most part to the propeller pitch which influences the engines' speed. So, a higher pitch leads to an increase of the engine speed, which has as effect an increase of the noise due to the bigger force in gear. Decreasing the pitch leads to a need of lower force from the free engine which leads to a lower engine speed and to a lower noise level.

To observe tonal components and especially the values in the critical frequency domain for the human ear, a spectral analysis of the noise is required. Thus, the spectral analyses of the acoustic signals in each microphone are presented in the following figures. The acoustic signals during the flight were processed by using 1/3 octave band analysis resulting in an averaged spectrum for the entire flight period. Figure 10 presents the averaged noise spectra for each area. The averaged noise spectra for the areas 3 and 4 indicate that the amplitudes of all frequency bands are significantly higher than the other areas, especially in the frequency domain of 400–5000 Hz. Under 400 Hz the noise is generated mainly by the vibration of the helicopter structure and due to the big wavelengths, the noise differences between the areas are small.

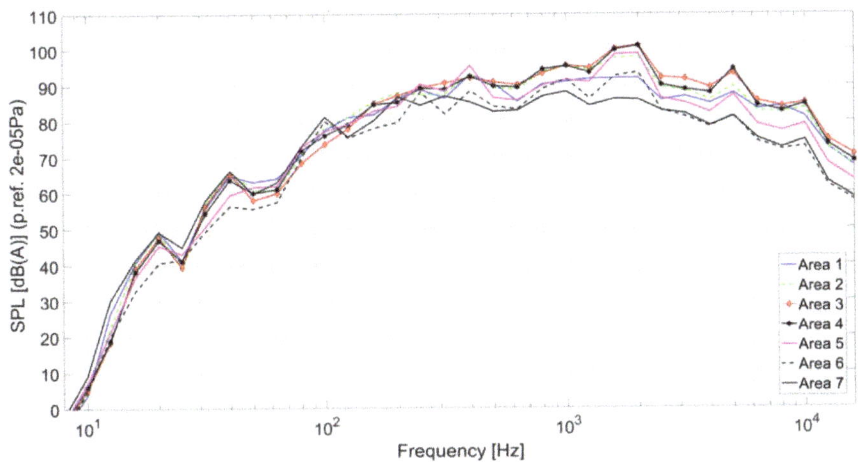

Figure 10. Averaged noise spectra in 1/3 octave bands for each measured area.

Over the frequency of 400 Hz differences between the areas start to be higher due to the fact that the acoustic sources are generated locally.

Table 2 presents the overall noise level resulting from the acoustic spectra presented in Figure 10. The highest noise level is measured in Area 3, which is situated below the transmission gear, followed by the adjacent areas. It can be noticed that the noise level decreases by more than 9 dB to the back of the helicopter. The differences of the overall noise levels presented in Table 2 are highlighted using color code for each zone where the red one has high amplitude and the dark green is the lowest one.

Table 2. The averaged overall level for each measuring zone.

Zone	1	2	3	4	5	6	7
LAeq dB(A)	101.7	104.7	106.5	105.9	103.6	99.7	97.2

Based on the spectral analysis presented in Figure 10, the average sound spectra were calculated for the entire helicopter, by using Equation (1). The average sound spectrum of the entire helicopter, presented in Figure 11, reveals several noise peaks at 400 Hz, 1 kHz, 2 kHz and 5 kHz. It must be mentioned that the peak from 2 kHz, a frequency where the human ear has an amplification of +1.2 dB (spectral weighting adjustment factor must be applied when converting between the weightings) [25], leads to an increase of the acoustic discomfort.

Figure 12 presents the FFT (Fast Fourier Transform) spectral analysis corresponding to the time step when the noise in the cabin was the highest. This level was recorded in area 3 by microphone 9 when an overall level of 112 dB(A) was measured, and the biggest spectral amplitude was at 1.7 kHz component with a value of 104.7 dB(A).

For a better representation of the spectral components in time, Figure 13 presents a FFT analysis in the time domain for the entire flight time, for multiple noise signals. The time domain analysis has been performed with a spectral resolution $\Delta f = 25$ Hz, time step $t_s = 1$ s, data blocks overlap of 70%. The results are presented as sonograms where the X axis represents the frequency, the Y axis represents the time and the color codes represent the sound pressure level amplitude.

As can be observed from Figure 13a, at around 400 Hz two tonal components are identified which are generated by the speed of the two turbo shaft engines, the differences in frequency being produced by the different operating speeds of these two. The spectral components of the engines' speed have the highest amplitudes in microphone 1 because the engines are located above it.

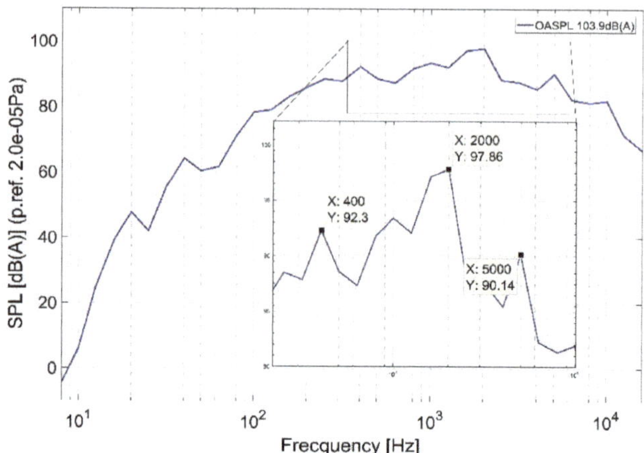

Figure 11. The average acoustic spectrum across the helicopter with maximum peak sounds at 400 Hz, 1 kHz, 2 kHz and 5 kHz.

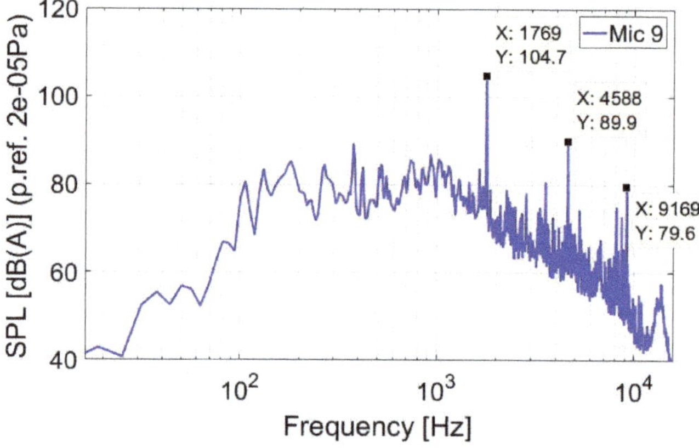

Figure 12. Instantaneous FFT spectrum of the highest noise level measured in microphone 9.

The signals of microphones 6 and 9 highlight a broadband noise in the frequency range of 150–2000 Hz with strong tonal components; these are generated by the gear. The analysis of the signal from microphone 16 highlights that the influence of the noise produced by the turbo engines and gear is considerably reduced; instead a tonal component is identified at the frequency of 100 Hz with 2 harmonics. This tonal component can correspond to the main rotor–tail rotor interaction noise.

5.2. Ground Results

5.2.1. Part 1. Emission–Reception

Figure 14 shows the variation in time of the sound pressure level for each microphone during the ground tests. It can be noticed that there was no variation of the noise and the small differences of the noise between the outside microphones indicate that a diffuse acoustic field was generated on the outside helicopter structure. Inside, the differences between the acoustic pressure levels are given by the acoustic modal behavior of the cabin enclosure, where in some places low amplitudes are recorded, in others high amplitude. The differences between the overall levels of the inside microphones indicate that some doors' weather-strips cannot assure a proper acoustic sealing so the door region is a weak point regarding the transmission of outside noise.

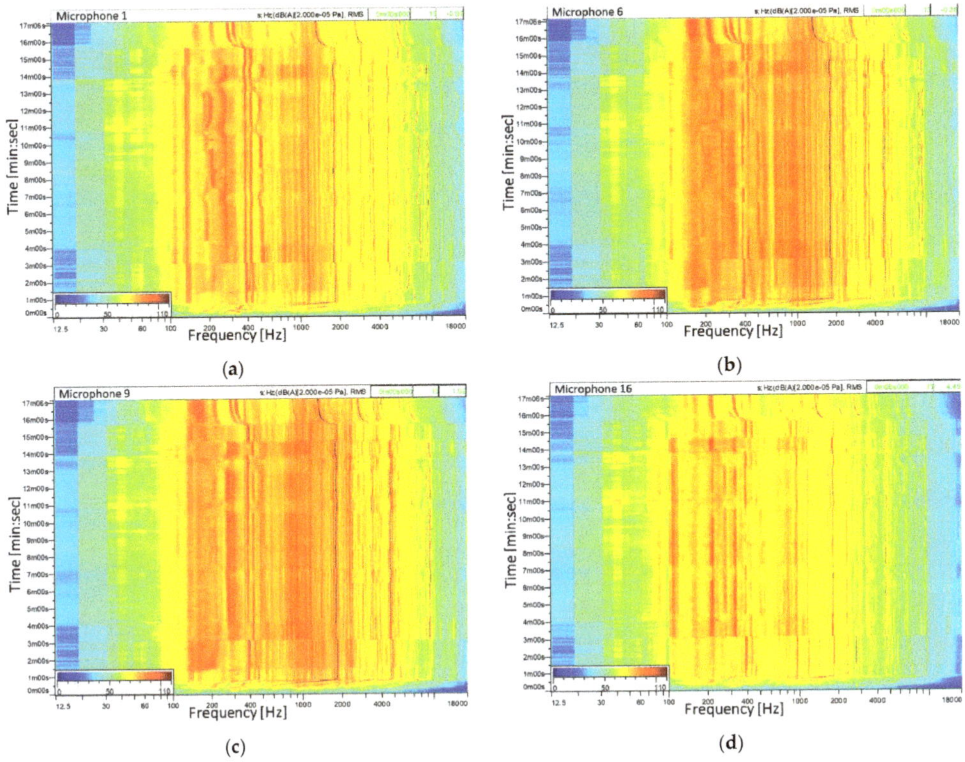

Figure 13. FFT analysis in time domain of the noise signals from: (**a**) microphone 1, (**b**) microphone 6, (**c**) microphone 9, (**d**) microphone 16.

The 1/3 octave band spectral analysis presented above in Figure 14b shows that in the 10-3000 Hz frequency domain, the maximum difference between the acoustic spectrums is 3 dB. The difference between the maximum and the minimum of the spectral values increases after 3000 Hz to a maximum value of 5 dB, given by the acoustic directivity pattern of the loudspeakers, directivity which presents acoustic side lobes. Using the spectral analysis of the inside and outside microphones, two averaged spectra resulted which correspond to the outside (emission) and inside (reception) noise levels, Figure 14b.

Based on the average pressure measured outside and inside of the helicopter, the acoustic insertion loss of the helicopter structure together with the IAR soundproofing structure (Figure 15) (IL) was computed for which an overall attenuation of 28 dB was obtained. These results can be used in a future analysis when a new soundproofing material will be implemented on the helicopter. The IL curve presents peaks and valleys that are produced by the acoustic modal response of the entire cabin; at low frequency the insulation is weak due to the thin frame of the helicopter and at high frequencies the insulation increases up to 35 dB at high frequencies.

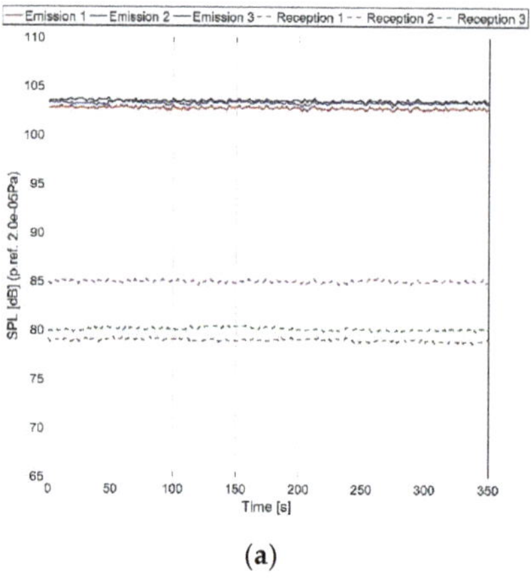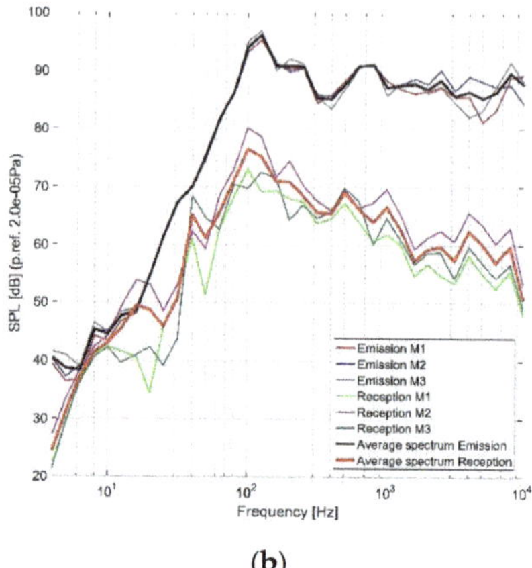

(a) **(b)**

Figure 14. (**a**) Variation of the noise level during the measurements, (**b**) spectral analysis 1/3 octave during measurements inside and outside the helicopter.

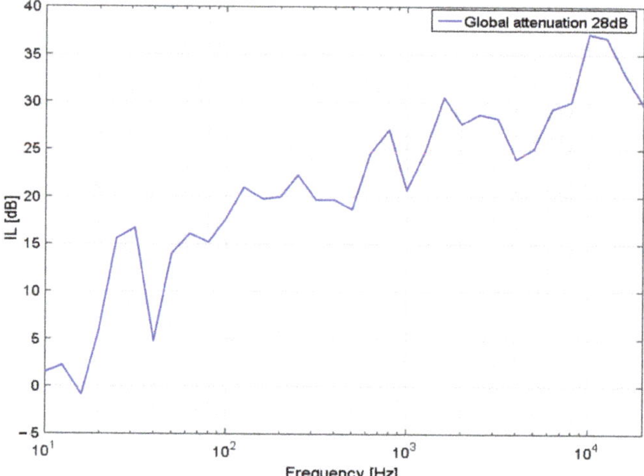

Figure 15. IL of helicopter structure with mounted soundproofing solution.

5.2.2. Part 2. Acoustic Intensity

The acoustic intensity measurements were carried out according to SR EN ISO 9614-2:2002 [22]. As presented in Figure 6, the used measurement grid included the surface of the helicopter's door, as well as its frame. During the measurements, a pink noise was generated, using the same configuration for the loudspeakers as in the IL determination. The results highlight the areas where acoustic leaks are located (Figure 16), which are represented with dark red. Considering the psychoacoustic factor and the A weighting, in Figure 16a one can observe that most leaks are situated in the upper part of the door, the part where the engine exhaust is situated. So, during the engines' functioning the exhaust noise is propagating into the cabin through this region.

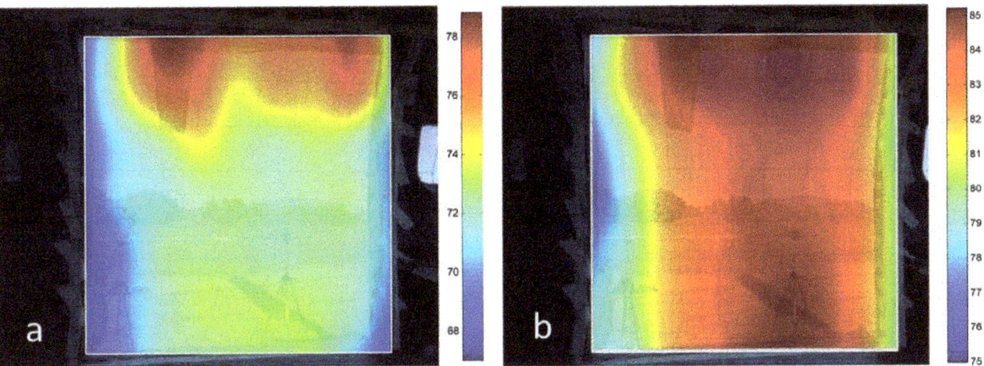

Figure 16. Acoustic intensity over the entire surface of the helicopter's door: (**a**) A weighted values (dB(A)), (**b**) linear values (dB).

6. Conclusions

This paper presents a complete acoustic evaluation for the IAR PUMA 330 helicopters with in-flight tests and tests performed on the ground.

One innovative aspect of this paper, compared to other similar articles, is the implementation of a procedure for evaluating the sound insulation of the helicopter structure using a set of speakers located outside the helicopter that generates a diffuse sound field. The second important aspect of this study is the identification of acoustic leaks in the area of the door sealing elements.

The acoustic measurements performed during the flight highlighted that in the area situated under the transmission gear, higher noise levels were measured. The noise levels decrease by 4.8 dB in the pilots' cabin and in the other direction noise decreases by 9.3 dB(A). Thus, it is noted that the sound pressure levels during flight vary from 97.2 dB(A) on the helicopter tail to 106.5 dB(A) under the transmission gear. Based on all equivalent sound pressure levels the average sound level was computed, having a peak value of 106 dB at 2 kHz. The correlation between the flight data and the noise shows that the pitch is the main parameter that influences the inside noise by almost 5 dB(A). The ground test highlighted that the acoustic leaks are caused by the doors' weather-strips and this must be changed or a new door seal system must be designed. Considering that the noise levels are different inside the helicopter, different soundproofing structures can be used to optimize the total mass of these structures. Under the transmission gear a heavier acoustic soundproofing material can be used to obtain a higher attenuation and in other areas, except the back of the helicopter where because of the existing low noise there is no need for acoustic treatment, a lighter structure can be used in comparison with the one used.

Inside acoustic evaluation helicopter measurements are quite rare in the literature. This paper can also provide an available data set and reference for researchers in further investigations.

Author Contributions: Conceptualization, G.C., A.-C.T., L.I.D.; software, M.D.; experimental analysis, M.D., G.C., A.-C.T., L.I.D.; supervision, G.C.; project administration, G.C. All authors have read and agreed to the published version of the manuscript.

Funding: This study was funded by Romanian Ministry of Research and Innovation Grant no. 97BG/2016.

Institutional Review Board Statement: Not applicable.

Informed Consent Statement: Not applicable.

Data Availability Statement: The datasets used and analyzed during the current study are available from the corresponding author on request.

Conflicts of Interest: The authors declare no conflict of interest.

References

1. Eurpoean Commission. *Aeronautics and Air Transport: Beyond Vision 2020 (Towards 2050) A Background Document from ACARE (The Advisory Council for Aeronautics Research in Europe)*; Eurpoean Commission: Brussels, Belgium, 2010.
2. Pathak, K. Noise Level Monitoring. *Environ. Monit.* **2009**, *I*, 247–250.
3. Schmitz, F.H. Chapter 2: Rotor Noise. In *Aeroacoustics of Flight Vehicles, Theory and Practice*; Hubbard, H.H., Ed.; Acoustical Society of America through the American Institute of Physics: College Park, MD, USA, 1995; Volume 1.
4. Schmiz, F.H.; Yu, Y.H. Helicopter impulsive noise: Theoretical and experimental status. *J. Sound Vib.* **1986**, *109*, 361–422. [CrossRef]
5. Miller, R.H. Methods for rotor aerodynamic and dynamic analysis. *Prog. Aerosp. Sci.* **1985**, *22*, 13–160. [CrossRef]
6. Edwards, B. Psychoacoustic Testing of Modulated Blade Spacing for Main Rotors. NASA/CR-2002-211651. Available online: https://www.cs.odu.edu/~{}mln/ltrs-pdfs/NASA-2002-cr211651.pdf (accessed on 6 August 2021).
7. Leverton, J.W. Helicopter Noise: What is the Problem? *Vertiflite* **2014**, *60*, 12–15.
8. Magliozzi, B.; Metzger, F.B.; Bausch, W.; King, R.J. *A Comprehensive Review of Helicopter Noise Literature*; Report No. FAA-RD-75-79; National Technical Information Service: Springfield, VA, USA, 1975; pp. 13–34.
9. AgustaWestland, T.; Swisshelicopter, M. Helicopter Noise Reduction Technology (Status Report). Available online: https://www.icao.int/environmental-protection/Documents/Helicopter_Noise_Reduction_Technology_Status_Report_April_2015.pdf (accessed on 24 May 2021).
10. Küpper, T.E.; Steffgen, J.; Jansing, P. Noise Exposure During Alpine Helicopter Rescue Operations. *Annals Occup. Hyg.* **2004**, *48*, 475–481. [CrossRef]
11. Julien, C.; Franck, M.; Yannick, U.; Aubourg, P.A. Comprehensive approach for noise reduction in helicopter cabins. In Proceedings of the 35th European Rotorcraft Forum, Hamburg, Germany, 22–25 September 2009.
12. Caillet, J.; Marrot, F.; Dupont, P.; Malburet, F.; Carmona, J.C. Nearfield Acoustical Holography measurement inside a helicopter cabine. In Proceedings of the European Test & Telemetry Conference, Toulouse, France, 7–9 June 2005.
13. Marze, H.J.; Dambra, F.N. Helicopter internal noise treatment. Recent methodologies and practical applications. In Proceedings of the 11th European Rotorcraft Forum, London, UK, 10–13 September 1985.
14. Schmitz, F.H.; Greenwood, E.; Sickenberger, R.D.; Gopalan, G.; Sim, B.W.C.; Conner, D.; Moralez, E., III; Decker, W.A. Measurement and Characterization of Helicopter Noise in Steady-State and Maneuvering Flight. In Proceedings of the American Helicopter Society Annual Forum, Virginia Beach, VA, USA, 1–3 May 2007.
15. Novak, D.; Bucak, T.; Miljković, D. Comparative helicopter noise analysis in static and in-flight conditions. *J. Acoust. Soc. Am.* **2008**, *123*, 3537. [CrossRef]
16. Tatić, B.; Bogojević, N.; Todosijević, S.; Šoškić, Z. Analysis of noise level generated by helicopters with various numbers of blades in the main rotor. In Proceedings of the 23th National Conference & 4th International Conference Noise and Vibration, Nis, Serbia, 17–19 October 2012.
17. Nelson, G.; Rajamani, R.; Erdman, A. Noise control challenges for auscultation on medical evacuation helicopters. *Appl. Acoust.* **2014**, *80*, 68–78. [CrossRef]
18. Caillet, J.; Marrot, F.; Unia, Y.; Aubourg, P.-A. Comprehensive approach for noise reduction in helicopter cabins. *Aerosp. Sci. Technol.* **2012**, *23*, 17–25. [CrossRef]
19. Available online: www.iar.ro/en/ (accessed on 24 May 2021).
20. Available online: www.easa.europa.eu (accessed on 24 May 2021).
21. ISO 5129 Acoustics—Measurement of Sound Pressure Levels in the Interior of Aircraft During Flight. Available online: https://www.iso.org/standard/32344.html (accessed on 24 May 2021).
22. SR EN ISO 9614-2/2000 Acoustics—Determination of Sound Power Levels of Noise Sources Using Sound Intensity. Available online: https://www.iso.org/standard/21247.html (accessed on 24 May 2021).
23. Available online: http://www.comoti.ro/en/laboratoare_experimentari_4_2.htm (accessed on 24 May 2021).
24. ISO 3746 Acoustics—Determination of Sound Power Levels and Sound Energy Levels of Noise Sources Using Sound Pressure—Survey Method Using an Enveloping Measurement Surface Over a Reflecting Plane. Available online: https://www.iso.org/standard/52056.html (accessed on 24 May 2021).
25. IEC61672: 2014 Standard: Electroacoustics—Sound Level Meters. Available online: https://standards.iteh.ai/catalog/standards/sist/152936bc-bdc1-486a-93fb-fe9b8fb6526f/sist-en-61672-1-2014 (accessed on 24 May 2021).

Article

Field Study of the Interior Noise and Vibration of a Metro Vehicle Running on a Viaduct: A Case Study in Guangzhou

Lei Yan [1,2], Zhou Chen [1], Yunfeng Zou [1,2], Xuhui He [1,2], Chenzhi Cai [1,2,*], Kehui Yu [1] and Xiaojie Zhu [3]

[1] School of Civil Engineering, Central South University, Changsha 410075, China; leiyan@csu.edu.cn (L.Y.); zhouchen520@csu.edu.cn (Z.C.); yunfengzou@csu.edu.cn (Y.Z.); xuhuihe@csu.edu.cn (X.H.); 164801022@csu.edu.cn (K.Y.)
[2] National Engineering Laboratory for High Speed Railway Construction, Changsha 410075, China
[3] Guangzhou Metro Corporation, Ltd., Guangzhou 510000, China; zhuxiaojie@gzmtr.com
*   Correspondence: chenzhi.cai@csu.edu.cn

Received: 11 March 2020; Accepted: 16 April 2020; Published: 19 April 2020

Abstract: The interior noise and vibration of metro vehicles have been the subject of increasing concern in recent years with the development of the urban metro systems. However, there still is a lack of experimental studies regarding the interior noise and vibration of metro vehicles. Therefore, overnight field experiments of the interior noise and vibration of a standard B-type metro train running on a viaduct were conducted on metro line 14 of Guangzhou (China). Both the A-weighted sound pressure level and linear sound pressure level were used to evaluate the interior noise signals in order to revel the underestimation of the low-frequency noise components. The results show that the interior noise concentrates in the low-to-middle frequency range. Increasing train speeds have significant effects on the sound pressure level inside the vehicle. However, two obvious frequency ranges (125–250 Hz and 400–1000 Hz) with respective corresponding center frequencies (160 Hz and 800 Hz) of the interior noise are nearly independent of train speed. The spectrum analysis of the vehicle body vibration shows that the frequency peak of the floor corresponds to the first frequency peak of the interior noise spectrum. There are two frequency peaks around 40 Hz and 160 Hz of the sidewall's acceleration level. The frequency peaks of the acceleration level are also independent of the train speeds. It hopes that the field measurements in this paper can provide a data set for researchers for further investigations and can contribute to the countermeasures for reducing interior noise and vibration of a metro vehicle.

Keywords: sound pressure level; field measurements; spectrum analysis; interior noise and vibration of vehicle

---

1. Introduction

In recent decades, the urban metro system has seen rapid development in eastern countries due to its fast speed, high efficiency, comfort, and environmental benefits, especially in China. According to recent statistics, the total length of China's operating metro lines has increased from 112 km among three cities in 2000 to about 6000 km among 41 cities in 2019. The elevated metro has become an alternative rather than the general underground metro type due to its low cost and short construction period, especially in a metropolis with a well-developed metro network, such as Beijing, Shanghai, and Guangzhou. The total length of operated elevated metros in China accounts for nearly a quarter of the whole 6000 km [1]. Although many benefits come from urban rail transit, the rail transit system has been considered the second greatest noise source affecting human modern lifestyles [2–4], after road traffic [5–7], but before airports [8,9], industries and wind turbines [10,11], and port activities [12,13]. Moreover, there are growing complaints about the noise inside the trains, as more passengers and metro staff will be exposed to interior train noise for a longer time with the expansion of the metro system [14,15]. According to statements released by the World Health Organization (WHO), noise exposure to 85 dB A-weighted over 45 min will lead to noise-induced hearing loss (NIHL) [16]. Meanwhile, recent investigations reveal that the accumulation of short-term noise exposure can also cause NIHL [17]. Furthermore, sleep disorders with awakenings, learning impairment, hypertension, ischemic heart disease and especially annoyance are the most common negative health effect related to prolonged exposure [18–21].

The interior noise inside a railway vehicle is composed of air-borne and structure-borne sound generated by exterior sources mainly including wheel–rail rolling effects, excitation phenomena due to the sleeper-passing frequency, and aerodynamic effects [22]. Many achievements concerning the interior noise problem of trains have been made theoretically and numerically. Eade and Hardy [23] investigated the transmission mechanisms of noise generated by various sources into a vehicle train through both the air-borne and structure-borne path. The spectrum results of interior noise show that low-frequency noise accounted for the largest proportion. Forssén et al. [24] proposed a statistical energy analysis model to predict the interior sound field of a railway vehicle. The prediction results were validated by a ray tracing method and scale model measurements. Zheng et al. [25] established a full-spectrum prediction method of the interior noise of high-speed train by considering both the air-borne and structural-borne noise. Dai et al. [26] presented a prediction method by applying statistical vibration and acoustic energy flow to obtain the full-spectrum interior noise of high-speed railway vehicles. Shi [27] employed a dynamic train-track interaction model, a finite element model, and an acoustic boundary element model to predict the interior noise of a high-speed vehicle at a speed of 200 km/h.

In addition to the aforementioned theoretical and numerical investigations, many field measurements have also been conducted. Han et al. [28] investigated the effects of rail corrugation on the interior noise and vibration of a metro vehicle based on the measurements. Li et al. [29] conducted field measurements to examine the interior noise and vibration of a railway vehicle at different speeds with respect to two different rail fastener stiffnesses. Fan et al. [30] analyzed the major interior noise sources and their corresponding transmission path into a high-speed vehicle through acoustic-vibration measurements as the train travels. Thompson [22] concluded that the low-frequency noise in a trail vehicle were the results of the structure-borne noise, while the air-borne noise contributed to the high-frequency noise of the train's interior noise according to measured results in British coaches. Conventionally, the A-weighted sound pressure level (SPL) has been widely used to evaluate the acoustical environment inside the train, and it has already been incorporated into some standards [31,32]. However, acoustic comfort inside the train cannot be achieved even when the A-weighted SPL inside the train meets the requirements of these standards due to the underestimated impacts of the low-frequency components of interior noise on people in the A-weighted SPL evaluation [33,34]. Soeta et al. [35] pointed out that the improvement of the acoustic environment in a train's carriages only through the reduction of the A-weighted SPL inside the carriage was impossible. The subjective felling should be considered in the evaluation of acceptable interior noise inside a train vehicle [22,36].

As there remains great concern regarding the acoustic environment of a metro vehicle, the lack of experimental studies about the interior noise and vibration of a metro vehicle, let alone investigations concerning a metro vehicle running on a viaduct, is noticeable. Moreover, with the trend of lighter trains and higher speeds, the increase of both air-borne and structure-borne sound will lead to the deterioration of the acoustic environment of a metro vehicle. This paper aims to determine the characteristics of the interior noise and vibration of a metro vehicle running on a viaduct. The measurements of vibration and noise were conducted in the standard B-type metro train on metro line 14 in Guangzhou (China). In order to analyze the underestimated effects of low-frequency noise components inside the vehicle, both the A-weighted SPL and linear SPL were adopted to evaluate the measured interior noise signals. The relationships between the interior noise and the vibration of the metro vehicle under various train speeds were investigated. The main noise sources of the metro vehicle were identified through the spectrum analysis of the measured vibration acceleration signals. The field measurements here can provide an available data set for researchers for further investigations, and could contribute to countermeasures for reducing interior noise and vibration of a metro vehicle.

2. Description of the Measurements

The main sources of interior noise can propagate by both air-borne and structure-borne paths, which includes rolling noise related to wheel and track characteristics and roughness and aerodynamic noise related to train speed and train type. The coupling effects of the train-track-bridge system may also affect the sources of interior noise when the metro vehicle is running on a viaduct [22]. The data analysis methods and the detailed measurements procedures are described here.

2.1. Data Analysis Methods

The total sound pressure level is a simple and direct parameter to quantify the sound level, which is defined in terms of the time-varying sound pressure level, as follows:

$$SPL = 10\log_{10}[\frac{1}{T}\int_0^T 10^{L(t)/10}dt]$$ (1)

where T is the required period of time and $L(t)$ is the sound pressure level at time t. This descriptor is a linear (i.e., unweighted) single value of sound level. In order to approximate the response of the human ear at low sound levels, the A-weighted total sound pressure level has been proposed and has been adopted in some international and Chinese standards. The A-weighted sound pressure level is defined as

$$SPL_A = 10\log_{10}[\frac{1}{T}\int_0^T 10^{L_{PA}(t)/10}dt]$$ (2)

where $L_{PA}(t)$ is the A-weighted sound pressure level at time t according to the A-weighted circuit.

The fast Fourier transform (FFT) is one of the most important numerical algorithms and has been widely used in signal analysis. The original signals can be transformed from the time domain to the frequency domain by adopting the FFT. Then, the spectral characteristics of the signals can be extracted. The FFT could be expressed as [37]

$$X(k) = \sum_{n=0}^{N-1} x(n)W_N^{kn}$$ (3)

where N represents the order number of each harmonic component, $x(n)$ represents the generic harmonic component as a complex number, $W_N = e^{-j(2\pi k/N)}$ and N = length $[x(n)]$. The FFT can be used for both quantifications of a noise problem and a vibration problem. Generally, the standardized one-third-octave band is adopted in the noise spectra analysis in order to obtain more detailed information.

2.2. Rail Condition

Wheel–rail roughness, especially rail corrugation, is the main source of vibration excitation for the interior noise of the metro vehicle. One of the most effective and economic countermeasures to prevent rail corrugation is rail grinding, especially before the operation. The field measurement of Metro line 14 of Guangzhou was conducted two weeks before its operation, and the rails were pre-grinded before its operation, as shown in Figure 1. Therefore, the influence of rail corrugation on the internal noise and the vibration of the subway vehicles could be considered as invariant during the field measurement.

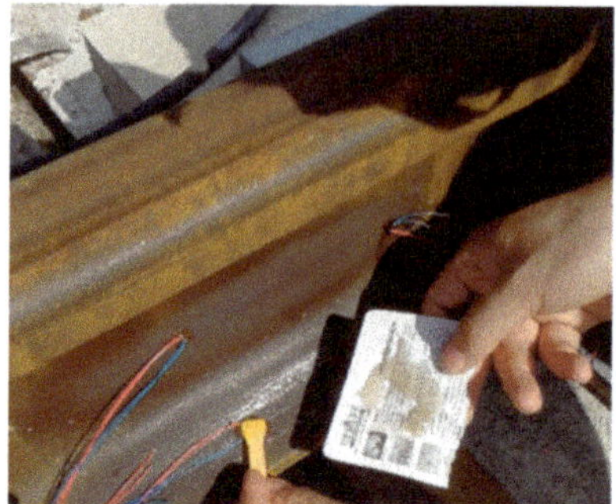

Figure 1. Photograph of rail condition.

2.3. Continuous Rigid-Frame Box-Girder Bridge

The running tests were conducted between overnight Dengcun station and Chicao station on metro line 14 in Guangzhou. The test section was an elevated metro line, which was mainly composed of 3×40 m concrete continuous rigid-frame box-girder bridge, as illustrated in Figure 2. The cross-sectional dimensions of the box girder are shown in Figure 3. The widths of the bridge deck and the bottom slab are 10 m and 2.4 m, respectively. The height of the box-girder is about 2 m. Two track systems are installed symmetrically on the bridge deck. The width of each track system is about 2.1 m. The distance between the central lines of the girder and the track system is 2.05 m.

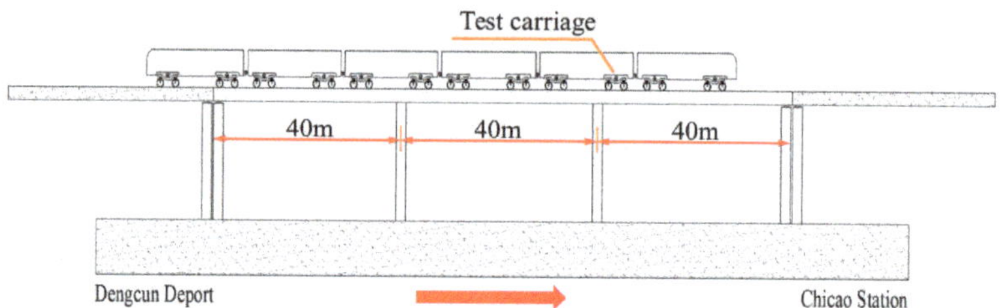

Figure 2. The schematic diagram of the running test.

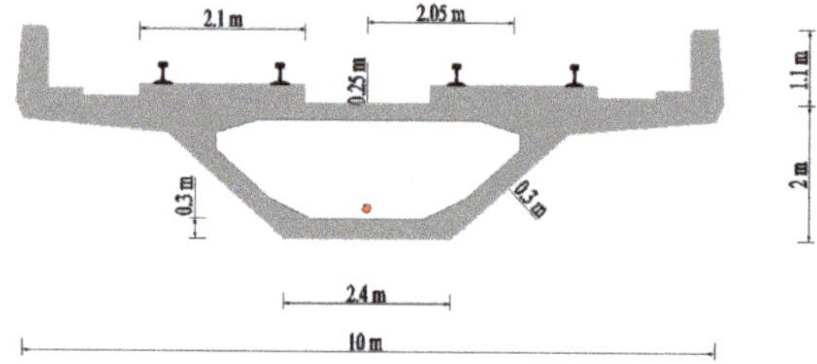

Figure 3. Schematic diagram of the box girder.

2.4. Metro Vehicle

The standard B-type metro train was used in the field test, which consists of four motor vehicles and two trailer vehicles, as illustrated in Figure 4. The head and trail are trailers, and 4 motor vehicles were arranged between the two trailer cars. The width and length of each vehicle are 2.8 m and 19.98 m, respectively. The maximum speed of the B-type metro train is designed to be 120 km/h. The measurement was conducted in the second passengers' carriage. Figures 5 and 6 demonstrate the installation positions of the microphones and accelerometers in the test vehicle. Eight microphones were placed along the centerline of the metro vehicle with distance at a distance of 2 m or 2.6 m, as illustrated in Figure 5. All of these microphones were 1.2 m above the floor according to GB 14892 [31] and ISO 3381 [32]. Two microphones labeled N1 and N8 were installed at both ends of the vehicle. Two microphones labeled N2 and N7 were placed near the top of the bogie frame. The acoustic signals inside the test vehicle were collected by a 24-bit intelligent acquisition and signal processing system (type: INV3020). Four vibration measuring points (V1 to V4) were arranged on the floor directly below microphones labeled N2, N4, N6, and N8, respectively, as shown in Figure 5. There were four other vibration measuring points (V5–V8) installed at the wall panel with the same height of the microphone corresponding to four floor measuring points, respectively. The vibration signals of the metro vehicle at different speeds were measured by the SDI Model 2210 accelerometers (4 mV/g sensitivity, ±2 g full scale). The test speed was controlled to be 20 km/h, 40 km/h, 50 km/h, 60 km/h, 80 km/h, or 115 km/h. A sound level calibrator was adopted to calibrate all the microphones before and after each running test, and the deviation between the calibration before and after the tests was less than 0.5 dB.

Figure 4. Photograph of the standard B-type metro train.

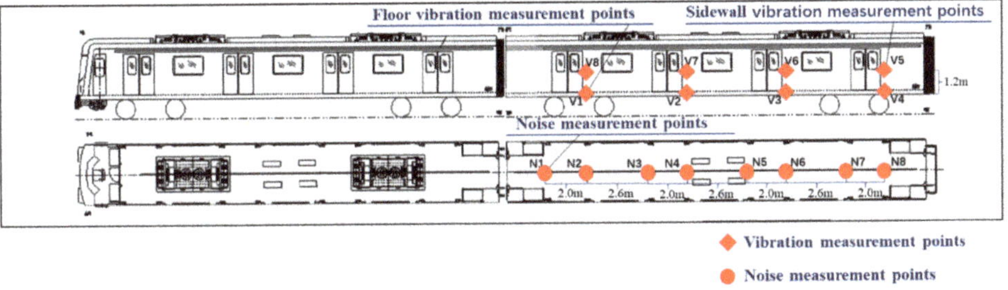

Figure 5. Layout of the noise and the vibration measurement points inside the vehicle.

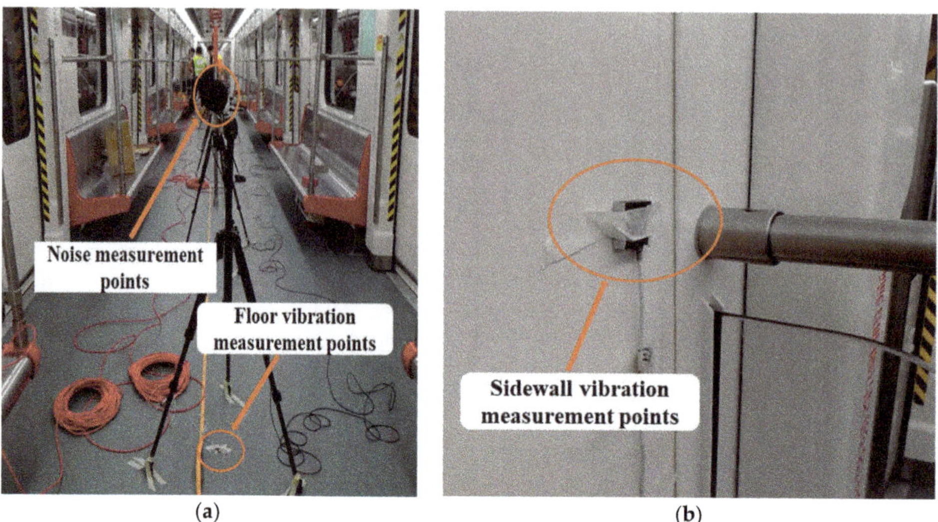

Figure 6. Photograph of equipment in the vehicle. (**a**) Noise measurement points and floor vibration measurement point; (**b**) sidewall vibration measurement point.

3. Results and Discussion

3.1. Interior Noise Spectra of the Metro Vehicle

According to the Noise Limit and Measurement for Trains of Urban Rail Transit standard [31] and referring to ISO 3381 standard [32], the noise measurement procedures satisfy the requirement that background noise is 10 dB lower than the interior noise. The A-weighted SPL has been widely used to evaluate the acoustic environment inside the train, and it has already been incorporated into some international and Chinese standards due to its consistency with the auditory characteristics of humans. The A-weighted SPL underestimates the noise below 1000 Hz, while, the energy of the noise inside the metro vehicle was mainly focused on the middle and low frequency regions, as shown in Figure 7. It can be seen from Figure 8 that the phenomenon of neglect of low frequency noise in the A-weighted SPL seems to be more obvious with the increase of train speed. Figure 9 shows the total A-weighted SPL of all the measuring points and its corresponding linear SPL under the train speed of 115 km/h. The acoustic comfort inside the train may not be achieved even though the A-weighted SPL inside the train meets the requirements of these standards due to the underestimated impacts of the low-frequency component of the interior noise on people in the A-weighted SPL evaluation. Therefore, both the linear SPL and A-weighted SPL were adopted here in order to reveal the effects of A-weighted SPL's correction in the low frequency range on the interior noise.

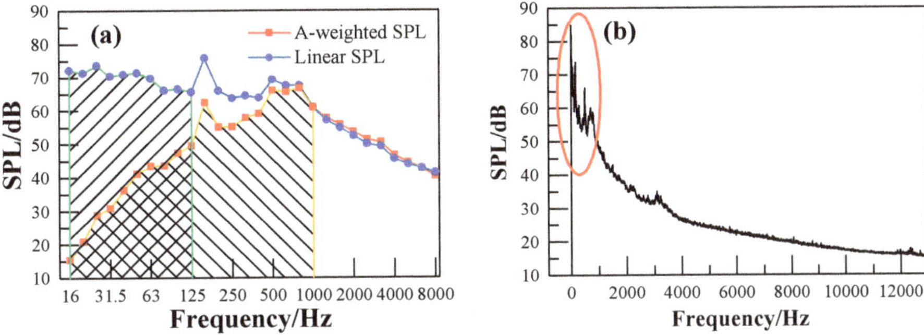

Figure 7. Interior noise at point N1 under the train speed of 50km/h: (**a**) one-third-octave band spectrum; (**b**) FFT spectrum.

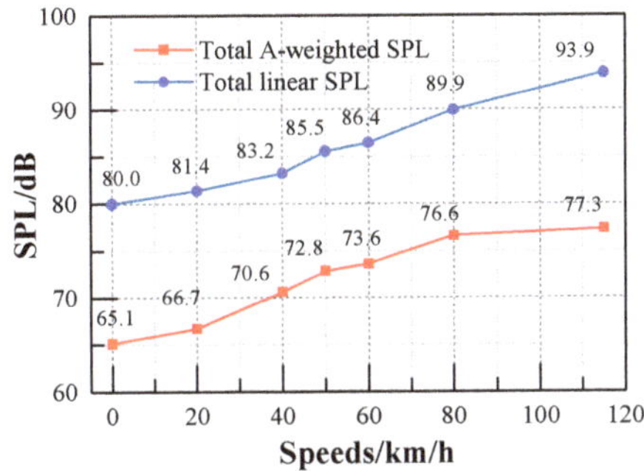

Figure 8. The total sound pressure levels at point N1 in respect of different train speeds.

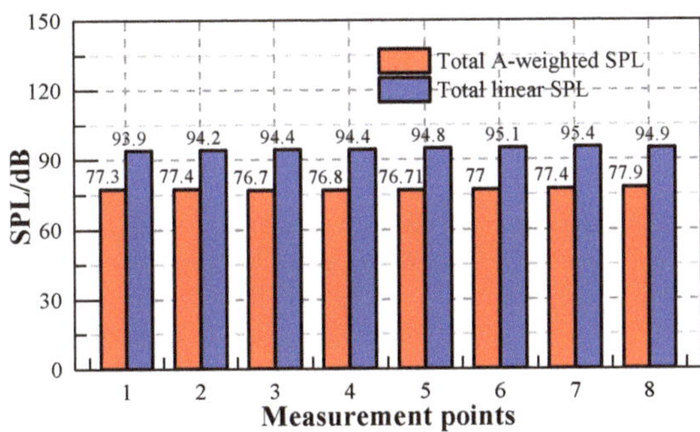

Figure 9. Comparison of total sound pressure levels using different methods with respect to different measuring points under the train speed of 115 km/h.

Since the combined effects of noise level and its frequency characteristics determines the acoustic environment of the vehicle, the total SPL and the spectrum characteristics of the interior noise are analyzed in this paper. The total SPL of each measuring point inside the vehicle with respect to different speeds is illustrated in Tables 1 and 2. It should be noted that 0 km/h represents the interior noise of the test vehicle when it is stationary with the operation of the auxiliary equipment in the vehicle. The A-weighted SPL of the elevated metro vehicle cannot exceed the limitation of 75 dBA [31]. It can be seen from Table 2 that the A-weighted SPL meets the limit requirements of no more than 75 dBA when the vehicle speed is less than 70 km/h. When the vehicle speed exceeds 80 km/h, the A-weighted SPL does not satisfy the requirement. It indicates that countermeasures should be taken. The A-weighted SPL increases with the increasing of the train speed. Similar results of Figure 8 can also be found regarding other measuring points according to the results illustrated in Tables 1 and 2. It can also be observed that the total SPL of all these measurement points has little difference under the same speeds. This indicates the uniform distribution of the interior noise inside the vehicle. Table 1 also show that the SPL is above 85 dB when the train speed reaches 50 km/h. Generally, long-term exposure to noise above 85 dB may damage hearing [16]. Therefore, noise reduction countermeasures are necessary during the operation of a metro line with a relatively high train speed.

Table 1. Total linear sound pressure level (SPL) of each measuring point with respect to different train speeds (dB).

Points	Speeds						
	0 km/h	20 km/h	40 km/h	50 km/h	60 km/h	80 km/h	115 km/h
N1	80.0	81.4	83.2	85.5	86.4	89.9	93.9
N2	79.5	80.6	82.9	85.1	86.2	89.8	94.2
N3	79.5	80.5	82.7	84.9	86.1	89.6	94.4
N4	79.4	80.5	82.7	84.9	86.1	89.5	94.4
N5	79.3	81.0	83.7	85.5	86.7	89.9	94.8
N6	80.0	81.3	83.7	85.8	86.9	90.2	95.1
N7	79.9	81.3	83.7	85.9	87.0	90.5	95.4
N8	79.1	81.0	83.9	85.7	86.7	90.1	94.9

Table 2. Total A-weighted SPL of each measuring point with respect to different train speeds (dBA).

Points	Speeds						
	0 km/h	20 km/h	40 km/h	50 km/h	60 km/h	80 km/h	115 km/h
N1	65.1	66.7	70.6	72.8	73.6	76.6	77.3
N2	65.1	66.2	70.6	72.6	73.4	76.4	77.4
N3	64.8	66.2	70.0	72.1	72.8	75.7	76.7
N4	64.8	66.4	70.3	72.2	72.9	76.0	76.8
N5	65.0	66.8	70.3	72.4	73.0	76.0	76.7
N6	64.9	66.9	70.3	72.5	72.9	76.1	77.0
N7	64.4	67.1	70.6	72.8	73.4	76.3	77.4
N8	64.0	67.8	70.8	73.3	73.9	76.9	77.9

Figures 10 and 11 show the interior noise spectrum corresponding to the train speeds of 50 km/h and 115 km/h, respectively. It can be seen from Figures 10a and 11a that the noise energy is mainly concentrated in the frequency range of 31.5–1000 Hz. There are two obvious peaks located at almost the same frequency in both interior noise spectra, as illustrated in Figure 10. A similar result can also be observed in Figure 11. However, it should be noted that the first peak in Figure 11b is not obvious. The first and second frequency peak in both Figures 10 and 11 center near 160 Hz and 800 Hz, respectively. This indicates that the interior noise of the metro vehicle is mainly composed of low and medium frequency components, and the characteristics of the interior noise spectrum are not related to the train speed. The first and second frequency peaks may be generated by the vibration of the vehicle itself and wheel–rail noise, respectively. As the train speed increases, the value of the first peak gradually decreases and tends to disappear, while the value of the second peak tends to increase. The wheel–rail noise may dominate in the interior noise of the metro vehicle as the train speed increases.

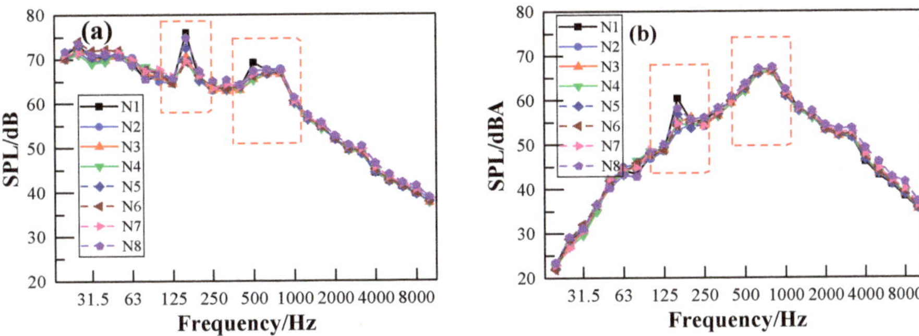

Figure 10. One-third-octave interior noise spectrum at different points (N1-N8) under the train speed of 50 km/h: (**a**) linear SPL; (**b**) A-weighted SPL.

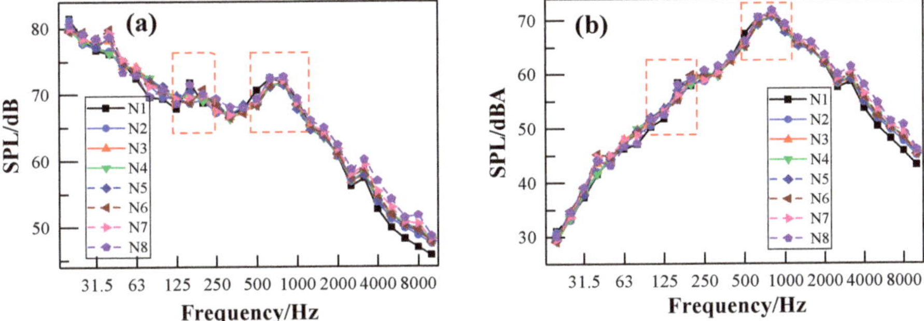

Figure 11. One-third-octave interior noise spectrum at different points (N1–N8) under the train speed of 115 km/h: (**a**) linear SPL; (**b**) A-weighted SPL.

Figure 12 shows the interior noise spectrum of N2 with respect to different train speeds. It can be observed that the results of SPL are nearly independent of train speed when the train speed is less than 20 km/h, especially above 500 Hz. This indicates that the interior noise is mainly caused by the auxiliary equipment of the vehicle when the train speed is under 20 km/h. Figure 13 also illustrates the dependence of the SPL on train speed. The SPL increases with increasing train speed. There are also two frequency peaks located around 160 Hz and 800 Hz, as shown in Figure 12. The frequency peaks are nearly independent of train speed. It is worth noting that the first frequency peak appears when the train speed is 50 km/h. In the high-frequency range, the interior noise at 115 km/h is at a relatively higher level than the interior noise at other train speeds, which may be due to the increase of aerodynamic noise. However, the curve of the noise spectrum decreases rapidly above 1250 Hz at these speeds due to the good sound insulation of the vehicle itself for high-frequency noise.

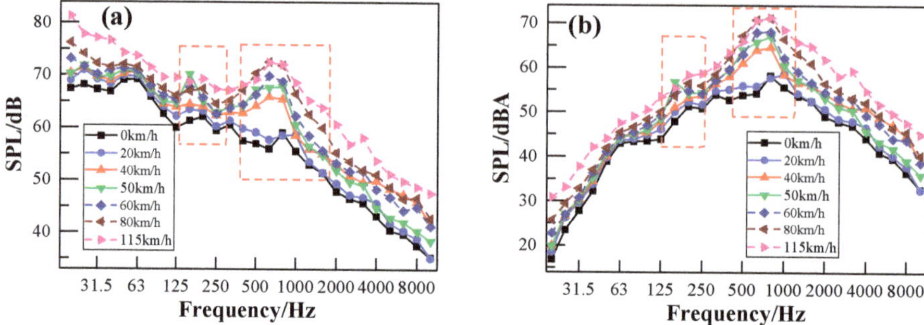

Figure 12. One-third-octave interior noise spectrum of point N2 under different train speeds: (**a**) linear SPL; (**b**) A-weighted SPL.

3.2. Vibration Spectra of the Metro Vehicle

The propagation path of noise in trains can be generally divided into the structural sound transmission and air sound transmission. The aim of vehicle vibration measurements is to reveal the relationship between the interior noise and the vibration of the metro vehicle. Figure 13a,b shows the vibration levels of the floor and sidewall, respectively, at a train speed of 50 km/h. Their corresponding frequency spectra using the FFT method are illustrated in Figure 14a,b, respectively. It can be observed that the vibration energy of the floor is mainly concentrated in the frequency range of 100–200 Hz. The frequency peak is located at 160 Hz, which corresponds to the first frequency peak of the interior noise spectrum in Figure 10. This means that the low-frequency noise may be generated by the vibration of the floor. Figures 13a and 14a show that the vibration peak of the measuring points nearing the bogie frame (V1 and V4) are higher than the measuring points in the middle of the vehicle (V2 and V3). This means that the vibration of the bogie frame caused by the wheel–rail excitation may be directly transmitted to the vehicle and stimulate the vibration of the floor. It can be seen in Figure 14a that the sidewall has one more obvious frequency peak around 40 Hz, apart from the same frequency peak as the floor. This kind of situation may be due to the secondary vibration of the sidewall caused by air sound. However, there is no significant difference in the vibration level of each measuring point around 40 Hz after applying the FFT transformation, as shown in Figure 14b. This may be the relatively low air-borne sound energy compared with the vibration energy of the bogie frame.

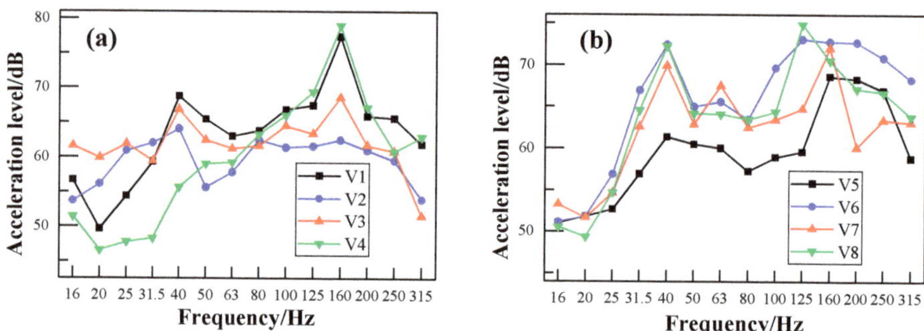

Figure 13. One-third-octave acceleration spectrum of different points at train speed of 50 km/h: (**a**) measuring points of the floor; (**b**) measuring points of the sidewall.

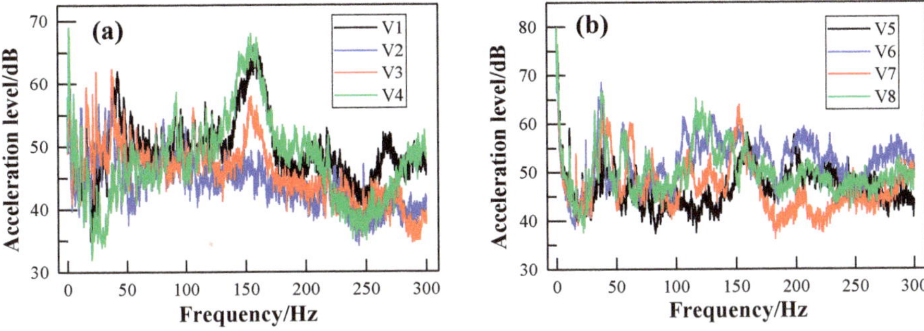

Figure 14. Fast Fourier transform (FFT) spectra of different points at train speed of 50 km/h: (**a**) measuring points of the floor; (**b**) measuring points of the sidewall.

Figure 15a,b shows the acceleration level of the floor measurement point V1 and the sidewall measurement point V8, respectively, with respect to different train speeds. Their corresponding frequency spectra using the FFT method are illustrated in Figure 16a,b, respectively. It can be seen from Figure 15 that both frequency peaks (around 40 Hz and 160 Hz) of the floor and sidewall are independent of the train speed. However, the vibration acceleration level increases with the increase of the train speed in both the low-frequency and high- frequency range. However, the vibration acceleration level increases with the increase of the train speed in both the low-frequency range and the high- frequency range. The acceleration level difference between the train speeds of 40 km/h and 80 km/h is not obvious. However, the acceleration level is much higher when the train speed reaches 115 km/h. This may be the instability of the vehicle itself at a relatively higher speed. Figure 16a shows that the acceleration levels of the floor around the second frequency peak decrease with the increase of the train speed, which may be the rolling effects of the rail. It can be seen from Figure 16b that the acceleration levels of the sidewall are nearly the same below 80 Hz. Furthermore, the acceleration levels between train speeds are also not obviously different above 80 Hz, except at 20 km/h. This indicates that the characteristics of the vehicle itself dominate the acceleration levels of the sidewall.

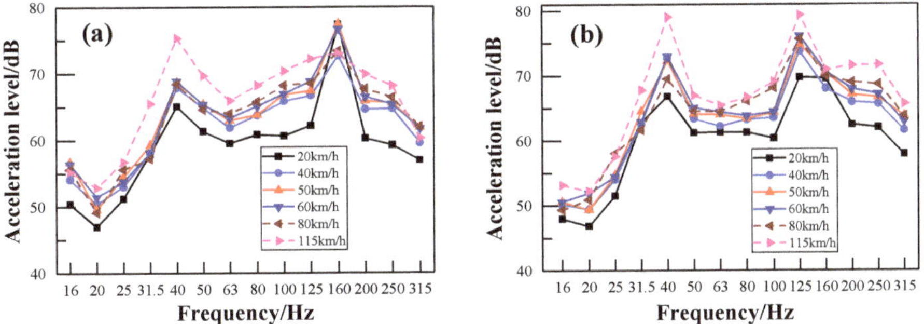

Figure 15. One-third-octave acceleration spectrum in respect of different train speeds: (**a**) measuring point V1 of the floor; (**b**) measuring point V8 of the sidewall.

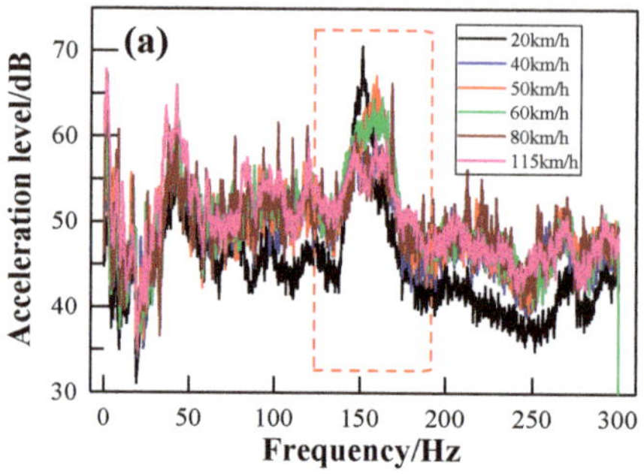
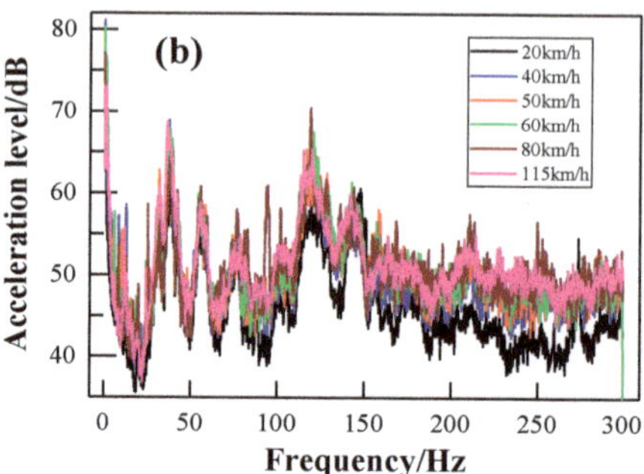

Figure 16. FFT spectra in respect of different train speeds: (**a**) measuring point V1 of the floor; (**b**) measuring point V8 of the sidewall.

4. Conclusions

Field experiments of the interior noise and vibration of a standard B-type metro train running on a viaduct were undertaken overnight on the Metro line 14 of Guangzhou (China). Both the A-weighted SPL and linear SPL were adopted to evaluate the measured interior noise signals. The FFT method was applied to measure vibrations of both the vehicle's floor and sidewall. The results show that the interior noise increases sharply with the increasing train running speed. However, the effects of the train's running speeds on the acceleration levels of the floor and sidewall are not apparent, especially in the range of 40–80 km/h. There are two obvious ranges (125–250 Hz and 400–1000 Hz) in the frequency domain of the interior noise. Their corresponding center frequencies are 160 Hz and 800 Hz, respectively. These two frequency peaks are nearly independent of train speed. The spectrum analysis of the vehicle body vibration shows that the frequency peak of the floor corresponds to the first frequency peak of the interior noise spectrum. This indicates that the vibration of the floor contributes to the low-frequency noise components of the interior noise. There are two frequency peaks of the sidewall's acceleration level, around 40 Hz and 160 Hz. The frequency peaks of the floor and sidewall are also independent of the train speed. This indicates that the characteristics of the vehicle itself dominate the frequency peaks of the acceleration levels of the floor and sidewall. The results show that different vibration reduction measures should be taken according to the characteristics of the floor and sidewall. Since field measurements of the interior noise and vibration of metro vehicles are quite rare in the literature, this paper can also provide an available data set and reference for researchers in further investigations.

Author Contributions: Conceptualization, L.Y. and C.C.; methodology, Y.Z. and X.H.; investigation, L.Y., Z.C., K.Y., and X.Z.; writing—original draft preparation, L.Y., Z.C., and C.C.; writing—review and editing, Y.Z. and X.H. All authors have read and agreed to the published version of the manuscript.

Funding: This research was funded by the National Key R&D Program of China (Grant No. 2017YFB1201204) and the Natural Science Foundation of China (Grant Nos. 51508580, U1934209 and U1534206).

Acknowledgments: The authors gratefully acknowledge the support of National Natural Science Foundations of China (Grant Nos. 51508580, U1934209 and U1534206), the National Key R&D Program of China (Grant No. 2017YFB1201204), and the Guangzhou Metro Group.

Conflicts of Interest: The authors declare no conflict of interest.

References

1. China Urban Rail Transit Association. Report on statistics and analysis of urban rail transit in 2018. *Urban Rail Transit* **2019**, *38*, 18–36.
2. Licitra, G.; Fredianelli, L.; Petri, D.; Vigotti, M.A. Annoyance evaluation due to overall railway noise and vibration in Pisa urban areas. *Sci. Total Environ.* **2016**, *568*, 1315–1325. [CrossRef]

3. Bunn, F.; Zannin, P.H.T. Assessment of railway noise in an urban setting. *Appl. Acoust.* **2016**, *104*, 16–23. [CrossRef]

4. Licitra, G.; Gallo, P.; Rossi, E.; Brambilla, G. A novel method to determine multiexposure priority indices tested for pisa action plan. *Appl. Acoust.* **2011**, *72*, 505–510. [CrossRef]

5. Cueto, J.L.; Petrovici, A.M.; Hernández, R.; Fernández, F. Analysis of the impact of bus signal priority on urban noise. *Acta Acust. United Acust.* **2017**, *103*, 561–573. [CrossRef]

6. Licitra, G.; Cerchiai, M.; Teti, L.; Ascari, E.; Bianco, F.; Chetoni, M. Performance Assessment of Low-Noise Road Surfaces in the Leopoldo Project: Comparison and Validation of Different Measurement Methods. *Coatings* **2015**, *5*, 3–25. [CrossRef]

7. Ruiz-Padillo, A.; Ruiz, D.P.; Torija, A.J.; Ramos-Ridao, Á. Selection of suitable alternatives to reduce the environmental impact of road traffic noise using a fuzzy multi-criteria decision model. *Environ. Impact Assess. Rev.* **2016**, *61*, 8–18. [CrossRef]

8. Gagliardi, P.; Teti, L.; Licitra, G. A statistical evaluation on flight operational characteristics affecting aircraft noise during take-off. *Appl. Acoust.* **2018**, *134*, 8–15. [CrossRef]

9. Iglesias-Merchan, C.; Diaz-Balteiro, L.; Solino, M. Transportation planning and quiet natural areas preservation: Aircraft overflights noise assessment in a national park. *Transp. Res. Part D Transp. Environ.* **2015**, *41*, 1–12. [CrossRef]

10. Michaud, D.S.; Feder, K.; Keith, S.E.; Voicescu, S.A.; Marro, L.; Than, J.; van den Berg, F. Exposure to wind turbine noise: Perceptual responses and reported health effects. *J. Acoust. Soc. Am.* **2016**, *139*, 1443–1454. [CrossRef] [PubMed]

11. Fredianelli, L.; Gallo, P.; Licitra, G.; Carpita, S. Analytical assessment of wind turbine noise impact at receiver by means of residual noise determination without the wind farm shutdown. *Noise Control Eng. J.* **2017**, *65*, 417–433. [CrossRef]

12. Bolognese, M.; Fidecaro, F.; Palazzuoli, D.; Licitra, G. Port noise and complaints in the north tyrrhenian sea and framework for remediation. *Environments* **2020**, *7*, 17. [CrossRef]

13. Fredianelli, L.; Nastasi, M.; Bernardini, M.; Fidecaro, F.; Licitra, G. Pass-by characterization of noise emitted by different categories of seagoing ships in ports. *Sustainability* **2020**, *12*, 1740. [CrossRef]

14. Gershon, R.R.M.; Qureshi, K.A.; Barrera, M.A.; Erwin, M.J.; Goldsmith, F. Health and safety hazards associated with subways: A review. *J. Urban Health* **2005**, *82*, 10–20. [CrossRef] [PubMed]

15. Münzel, T.; Sørensen, M.; Gori, T.; Schmidt, F.P.; Rao, X.; Brook, J.; Chen, L.C.; Brook, R.D.; Rajagopalan, S. Environmental stressors and cardio-metabolic disease: Part I-epidemiologic evidence supporting a role for noise and air pollution and effects of mitigation strategies. *Eur. Heart J.* **2017**, *38*, 550–556. [CrossRef]

16. Guski, R.; Schreckenberg, D.; Schuemer, R. WHO Environmental Noise Guidelines for the European Region: A Systematic Review on Environmental Noise and Annoyance. *Int. J. Environ. Res. Public Health* **2017**, *14*, 1539. [CrossRef]

17. Johnson, T.A.; Cooper, S.; Stamper, G.C.; Chertoff, M. Noise exposure questionnaire: A tool for quantifying annual noise exposure. *J. Am. Acad. Audiol.* **2017**, *28*, 14–35. [CrossRef]

18. Muzet, A. Environmental noise, sleep and health. *Sleep Med. Rev.* **2007**, *11*, 135–142. [CrossRef]

19. Dratva, J.; Phuleria, H.C.; Foraster, M.; Gaspoz, J.M.; Keidel, D.; Künzli, N.; Schindler, C. Transportation noise and blood pressure in a population-based sample of adults. *Environ. Health Perspect.* **2012**, *120*, 50–55. [CrossRef]

20. Babisch, W.; Beule, B.; Schust, M.; Kersten, N.; Ising, H. Traffic noise and risk of myocardial infarction. *Epidemiology* **2005**, *16*, 33–40. [CrossRef]

21. Bernardini, M.; Fredianelli, L.; Fidecaro, F.; Gagliardi, P.; Nastasi, M.; Licitra, G. Noise Assessment of Small Vessels for Action Planning in Canal Cities. *Environments* **2019**, *6*, 31. [CrossRef]

22. Thompson, D. *Railway Noise and Vibration: Mechanisms, Modelling and Means of Control*; Elsevier: Amsterdam, The Netherlands, 2008.

23. Eade, P.W.; Hardy, A.E.J. Railway vehicle internal noise. *J. Sound Vib.* **1977**, *51*, 403–415. [CrossRef]

24. Forssén, J.; Tober, S.; Corakci, A.C.; Frid, A.; Kropp, W. Modelling the interior sound field of a railway vehicle using statistical energy analysis. *Appl. Acoust.* **2012**, *73*, 307–311. [CrossRef]

25. Zheng, X.; Hao, Z.; Wang, X.; Mao, J. A full-spectrum analysis of high-speed train interior noise under multi-physical-field coupling excitations. *Mech. Syst. Signal Process.* **2016**, *75*, 525–543. [CrossRef]

26. Dai, W.Q.; Zheng, X.; Luo, L.; Hao, Z.Y.; Qiu, Y. Prediction of high-speed train full-spectrum interior noise using statistical vibration and acoustic energy flow. *Appl. Acoust.* **2019**, *145*, 205–219. [CrossRef]
27. Shi, Y.; Xiao, Y.G.; Kang, Z.C. Interior noise investigation for a passenger room of a high speed train under wheel-rail excitation. *J. Vib. Shock* **2009**, *28*, 95–98.
28. Han, J.; Xiao, X.B.; Wu, Y.; Wen, Z.F.; Zhao, G.T. Effect of rail corrugation on metro interior noise and its control. *Appl. Acoust.* **2018**, *130*, 63–70. [CrossRef]
29. Li, L.; Thompson, D.; Xie, Y.S.; Zhu, Q.; Luo, Y.Y.; Lei, Z.Y. Influence of rail fastener stiffness on railway vehicle interior noise. *Appl. Acoust.* **2019**, *145*, 69–81. [CrossRef]
30. Fan, R.P.; Su, Z.Q.; Meng, G.; He, C.C. Application of sound intensity and partial coherence to identify interior noise sources on the high speed train. *Mech. Syst. Signal Process.* **2014**, *46*, 481–493. [CrossRef]
31. *Noise Limit and Measurement for Train of Urban Rail Transit*; GB 14892; Standardization Administration: Beijing, China, 2012. (In Chinese)
32. *Railway Applications. Acoustics. Measurement of Noise Inside Railbound Vehicles*; ISO 3381; ISO: Geneva, Switzerland, 2011.
33. Persson, K.; Björkman, M. Annoyance due to low frequency noise and the use of the dB (A) scale. *J. Sound Vib.* **1988**, *127*, 491–497. [CrossRef]
34. Torija, A.J.; Flindell, I.H.; Self, R.H. Subjective dominance as a basis for selecting frequency weightings. *J. Acoust. Soc. Am.* **2016**, *140*, 843–854. [CrossRef] [PubMed]
35. Seidman, M.D.; Standring, R.T. Noise and quality of life. *Int. J. Environ. Res. Public Health* **2010**, *7*, 3730–3738. [CrossRef] [PubMed]
36. Ma, K.W.; Wong, H.M.; Mak, C.M. A systematic review of human perceptual dimensions of sound: Meta-analysis of semantic differential method applications to indoor and outdoor sounds. *Build. Environ.* **2018**, *133*, 123–150. [CrossRef]
37. Ning, J.; Lin, J.H.; Zhang, B. Time-frequency processing of track irregularities in high-speed train. *Mech. Syst. Signal Process.* **2016**, *66*, 339–348. [CrossRef]

International Journal of
Environmental Research and Public Health

Communication

Noise Annoyance in the UAE: A Twitter Case Study via a Data-Mining Approach

Andrew Peplow [1,*], Justin Thomas [2] and Aamna AlShehhi [3]

[1] Division of Engineering Acoustics, Department of Construction Sciences, Lund University, 221 00 Lund, Sweden
[2] College of Natural and Health Sciences, Zayed University, Abu Dhabi 144534, United Arab Emirates; justin.thomas@zu.ac.ae
[3] Electrical and Computer Engineering, Khalifa University, Abu Dhabi 127788, United Arab Emirates; aamna.alshehhi@ku.ac.ae
* Correspondence: andrew.peplow@construction.lth.se

Abstract: Noise pollution is a growing global public health concern. Among other issues, it has been linked with sleep disturbance, hearing functionality, increased blood pressure and heart disease. Individuals are increasingly using social media to express complaints and concerns about problematic noise sources. This behavior—using social media to post noise-related concerns—might help us better identify troublesome noise pollution hotspots, thereby enabling us to take corrective action. The present work is a concept case study exploring the use of social media data as a means of identifying and monitoring noise annoyance across the United Arab Emirates (UAE). We explored an extract of Twitter data for the UAE, comprising over eight million messages (tweets) sent during 2015. We employed a search algorithm to identify tweets concerned with noise annoyance and, where possible, we also extracted the exact location via Global Positioning System (GPS) coordinates) associated with specific messages/complaints. The identified noise complaints were organized in a digital database and analyzed according to three criteria: first, the main types of the noise source (music, human factors, transport infrastructures); second, exterior or interior noise source and finally, date and time of the report, with the location of the Twitter user. This study supports the idea that lexicon-based analyses of large social media datasets may prove to be a useful adjunct or as a complement to existing noise pollution identification and surveillance strategies.

Keywords: Twitter; noise; annoyance; geolocation; noise classification

Citation: Peplow, A.; Thomas, J.; AlShehhi, A. Noise Annoyance in the UAE: A Twitter Case Study via a Data-Mining Approach. *Int. J. Environ. Res. Public Health* **2021**, *18*, 2198. https://doi.org/10.3390/ijerph18042198

Academic Editors: Gaetano Licitra, Luca Fredianelli and Peter Lercher

Received: 30 December 2020
Accepted: 13 February 2021
Published: 23 February 2021

Publisher's Note: MDPI stays neutral with regard to jurisdictional claims in published maps and institutional affiliations.

1. Introduction

One of the striking technological patterns emerging at the end of the last century was the fast development and production of advanced devices (personal computers, smartphones and so on) across varying backgrounds [1]. One result of these improvements is the ascent of progressively enormous datasets, seemingly across all socioeconomic sectors. These datasets are often alluded to as "big data", a term generally utilized concerning the volume (speed of development) and an assortment of information. Beyond volume, velocity and value, commentators have also referred to the potential value attached to these datasets [1]. Value is the idea that industrially and culturally important data can possibly be utilized from these datasets for academic or commercial purposes. For example, Google search query data have been used for monitoring influenza outbreaks. Using the geolocation of search inquires, the spread of an outbreak can be monitored with greater speed and accuracy than conventional epidemiological surveillance techniques [2,3]. Psychological constructs, such as happiness or subjective wellbeing, have also been studied [4–6], self-concept [7], religiosity [8] and the use of profanity [9].

In the present study, we use similar big data analytic techniques to introduce a method for estimating both the prevalence and location of noise annoyance. In addition, the location and the number or tally of the tweet activity can be stored compared to frequency and geolocation, as discussed in the article [2]. In the last five years, the earliest known analytical methods for studying responses to sound from digital media was performed by the authors [10], who based their work on data from a repository of audio samples in Chatty Maps centered within London and Barcelona. The authors were successful in tagging and locating noise events and thus was able to create a large taxonomy and a lexicon reflecting both negative and positive aspects characterized into 6-noise sources; mechanical, transport, nature, human, music and indoor. This classification was also used in this work and by the authors [11,12], where the focus of their in-depth studies was based on noise response through social media, where a subset of Twitter responses (tweets) was analyzed in London, UK. The latter study grouped geographic areas into socioeconomics groups and was able to extract responses in terms of correlations to hypertension, which shows, above all, that meaningful conclusions can be drawn from these studies, which can benefit not only urban planners but also stakeholders in medical policies.

Noise is a growing public health concern, linked with issues that severely affect hearing, sleep and attribute to hypertension and heart disease. Individuals are increasingly using social media to voice complaints about problematic noise sources. This behavior—using social media to post noise-related complaints and comments—might help us better identify troublesome hotspots and take corrective action. This article is a concept study, which manages an examination of tweeted noise complaints sent from within the United Arab Emirates (UAE) during 2015. Such reports have been organized in an information base and grouped by (1) common types of noise sources (human factors, music, construction, traffic), (2) exterior or interior noise source (domestic, industrial, others such as ventilation noise) and (3) data and exact time of report with the location of the receptor/Twitter user.

A 2016 European study found that people living next to noisy roads were 25% more likely to have symptoms of depression than people in quieter areas, even when adjusting for socioeconomic factors, [13]. Noise criteria, for health reasons, are governed by sound energy levels averaged over a certain time period. The period is normally the 24 h cycle, which is divided into day/evening/nighttime (07:00–19:00–23:00–07:00) with weightings emphasizing the evening and nighttime levels. In 1995 the World Health Organization (WHO) declared, "*The main negative effects of such noise on people are disturbances of communication, rest and sleep, and general annoyance. Over long periods of time, these effects have a detrimental influence on wellbeing and perceived quality of life.*" [14].

Annoyance or irritation are commonly reported responses to ambient or environmental noise. Arising from non-positive effects on daily routines, thoughts, feelings, sleep deprivation, or daily rest can lead to negative emotions, such as distress, exhaustion, and other stressors [15–17]. Hence, this study is focused on the subjective response to noise, reporting annoyance by tweets. This has the advantage that the study is ecologically valid, capturing real-time complaints without any of the response biases or reactivity that can be associated with traditional self-report survey methodologies.

Noise pollution represents a complex issue in the evaluation of life equality, especially in built-up zones. Noise is defined as unwanted sound; it is the characteristic physical nature of sound that can transmit in the air and through building structures that represent both the level and character (for example, low-frequency sound through wall partitions) of noise annoyance. These sounds emanate from both predictable (traffic) or unpredictable (neighborhood) noise sources. Adjustment for bias, confounders, socioeconomic status (SES) and lifestyle habits are important factors to consider in scientifically controlled assessments on the impact of noise [18,19]. For example, in work in [20], the author has shown the negative impact on property prices due to traffic noise—these results can readily skew the response to a controlled questionnaire. It is therefore important to recognize the difficulty in considering lifestyle or biased opinion in scientific surveys.

In the WHO report, it was established that twice as many city-dwellers (23%) are reported as having suffered from noise compared to those living in rural settings (10%).

The document detailed reports largely from street or neighborhood noise, but nevertheless, the difference in numbers come as no surprise

The use of the Twitter dataset for quantifying noise disturbance will be enhanced by the availability of Geographical Positioning System (GPS) location data, as well as the day and actual instantaneous time in which the subject reported the "annoyance". This is valuable data, which has not been reported previously. The main aim of this cross-sectional pilot study is to assess the subjective noise annoyance and disturbance among population groups in or surrounding built apartments or villas situated within the emirates of UAE. To begin to comprehend the actual perception of unwanted sounds by residents, we present the analysis of reported complaints of noise pollution registered in the United Arab Emirates (UAE) via tweets. This methodology was implemented via open-access Python language, which has the capability to "tag" noise complaints via location, which could also be implemented as "Live" monitoring.

2. Methodology

The data used in this study was a randomly extracted subset of the UAE Twitter data for 2015 using their Historical PowerTrack enterprise product, although it should be possible to return similar results from a Twitter API. The dataset comprised 8.2 million tweets—approximately 10% of the total number of tweets that year—collected between 1 January 2015 to 31 December 2015. Provided by Twitter, the company, the material included as part of a large data download service established to support research. The data obtained included fields related to the user and fields related to the text (tweet). The data were collected via a Query search and coded using Python/Anaconda. This allowed the body text of each tweet to be subjected to query criteria, such as body text, which contains any one of the keywords from a chosen lexicon. The dataset was explored to check if it corresponded to expected national norms, for example, the percentage of tweets per emirate, the rate of Arabic use by Emirates. The data confirmed all expectations. Using a subset is, however, a limitation, and future studies should use larger datasets, ideally comprising the whole corpus of tweets for the region and timeframe under exploration. User features included display name and user description. Text level features included text language, geolocation, location name, and posted time. There are 24 different languages; 44% of tweets were Arabic and 39% English. For the purpose of this case study, we were limited to exploring English tweets only. Table 1 summarizes additional information concerning the data set.

Table 1. Breakdown of language use and unique users from the United Arab Emirates (UAE) Twitter dataset for 2015.

Language	Number of Tweets	Unique Users
Arabic	3,126,163	58,776
English	2,816,777	124,543
Other	2,262,602	6175
Total	8,205,542	189,494

Although the objectives of this article are factors involved in the study of "tweeting" the user's annoyance of noise, it is important to consider the layout of the data collected. The size of data, "participation patterns", and coverage, with details on individual cases and more specific patterns, are covered. To avoid missing descriptors, we identified words in British-English and American-English using parentheses and included many versions of words such as "noise" or "noisy" by wild-card descriptors, (*). We also decided against adopting the Arabic language due to complications; the lexicon, Table 2, was used to filter the data from which a total of 272 tweets were identified. The number of "hits" we were able to establish as related to a noise incident was crucially determined by the wording in the lexicon. After many attempts, convergence was not always certain; we decided on the lexicon shown below. Convergence here is meant in the sense of convergence in a reasonable time. This was not performed in a truly scientific manner but should be designed with more diligence in further attempts. Basically, we found the lexicon we used to provide the most efficient number of useful "hits". However, most of, which were false-positives, for example, *"Sleeping at Last's music is phenomenally, sensationally, and truly beautiful."* was sent on 18 April 2015. However, one example of an annoyed tweeter, *"Hey ya, construction noise from the site between Mag218 & 23 Marina is to [sic] loud "*, tweeted on 5 May 2015, was included. Manually removing these false positives reduced the dataset to 38 tweets positively identified as strongly correlated to the sender's annoyance. Data for the years 2016 and 2017 were available to the authors, but the material was incomplete or only partially available in some areas. To determine any trends over a full 12 month period, we decided to use the 2015 data exclusively.

Table 2. Lexicon used for filtering Tweets, UAE 2015 (*-represents Wildcard)

#	WORD	AND	AND	OR	OR	OR
1	neighbo(u)r	loud				
2	neighbo(u)r	rowdy				
3	neighbo(u)r	music	annoy *		disturb *	nerves
3	neighbo(u)r	nois *	annoy *		disturb *	nerves
4	music	loud	annoy *	too	disturb *	nerves
5	party	loud	annoy *	too	disturb *	nerves
6	construction	nois *	annoy *	too	sleep *	nerves
7	construction	loud				
8	construction	racket				
9	construction	sleep	annoy *	disturb *		nerves
10	people	shouting	next door	neighbor(u)r	disturb *	nerves
11	people	yelling	next door	neighbor(u)r	disturb *	nerves
12	people	screaming	next door	neighbor(u)r	disturb *	nerves
13	crowd	shouting	next door	neighbor(u)r	disturb *	nerves
14	crowd	yelling	next door	neighbor(u)r	disturb *	nerves
15	crowd	screaming	next door	neighbor(u)r	disturb *	nerves
16	hotel	noise	annoy *	too	disturb *	nerves
17	bar	noise	annoy *	too	disturb *	nerves
18	club	noise	annoy *	too	disturb *	nerves
19	airport	noise	annoy *	too	disturb *	nerves
20	plane	noise	annoy *	too	disturb *	nerves
21	traffic	noise	annoy *	too	disturb *	nerves
22	hotel	loud	too			
23	bar	loud	too			
24	club	loud	too			
25	airport	loud	too			
26	plane	loud	too			
27	traffic	loud	too			
28	traffic	sleep	annoy *	disturb *		nerves
29	jetski	nois *	annoy *	too	disturb *	nerves
30	dog	bark *	annoy *	disturb *		nerves
31	* plane	deafening	annoy *	disturb *		nerves

3. Results of Case Study

Sustained exposure to noise also has been correlated with cognitive impairment and behavioral problems in children, as well as the more obvious hearing damage and sleep deprivation. The European Environment Agency (EAA) has blamed 900 thousand cases of high blood pressure (hypertension), 43 thousand hospital admissions and 10 thousand cases of premature deaths a year in Europe on noise [21]. Road-traffic noise is the most pervasive noise: 125 million Europeans are exposed to sound pressure levels above 55 decibels (L_{den} 55)—considered as damaging to health. This value is calculated over day, evening and night periods with an emphasis on nighttime. The emphasis on nighttime exposure in L_{den} reflects the importance of sleep. Our data do not directly support this, but it could be concluded that most people are tired and are willing to tweet their dissatisfaction, Figure 1. However, there is a slight bias here to people who are predisposed to tweeting their emotions in a public forum. In addition, the possibility to tweet is only available to people who have access to this App. There may be a gap in the data, which corresponds to people, who work in noisy environments, but do not have access to the App to express their annoyance. Nevertheless, the figure shows that most tweets were reported late at night and in the early morning. Within 2015, 30% of noise annoyance tweets reported equally between October–December and January-March, and 20% equally between April–June and July–September.

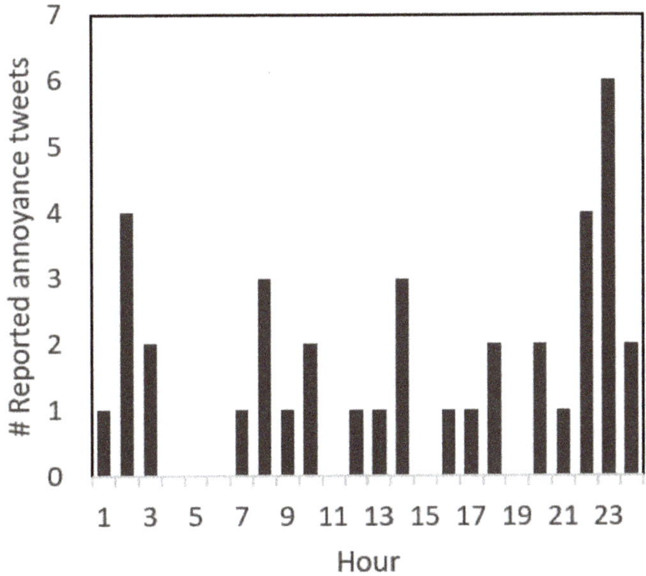

Figure 1. Frequency and time at which tweets were reported. Representing hourly intervals, the *x*-axis represents the 24 h period, i.e., midnight-to-midnight.

Due to network location availability, it was also possible to locate the emirate (in some cases the GPS coordinate in which the tweet was posted and, including the tweet, see Table 3) representing the total number of tweets 74% were in the Emirate of Dubai, 16% in Abu Dhabi, 8% Sharjah and the remaining of an unknown location. Of these, around 70% can be attributed to noise sources located within buildings. Unwanted noise from vehicles and airplanes is usually not categorized as a "noise nuisance", defined in the UK as "an unlawful interference with a person's use or enjoyment of land or some right over it, or in connection with it", [22]. As will be shown, residents tend to be more annoyed by noises that come from uncontrolled human sources (social interaction and increased volume music, Table 4) over predictable, controlled ones (road-traffic).

Table 3. Location of annoying noise tweets, UAE 2015.

Emirate	Number of Tweets	Location of Noise Source	Number of Tweets
Dubai	28	Exterior	12
Sharjah	3	Interior	21
Abu Dhabi	6	N/A	5
N/A	1		
Total	38		38

Table 4. Types of annoying noises reported by Twitter—UAE 2015.

Annoyance Source Type	#tweets
Music	25
Construction	4
Human	5
Traffic	2
Airplanes	1
Building services (e.g., air conditioning)	1
Total	38

The trend of complaints according to the source type activity is illustrated in Figure 2. Here we can see that music is the most common relative "offender". This contrasts with conclusions reported by [10] in, which they found degree the highest degree of annoyance was due to aircraft noise (60%), then road traffic (44%), neighborhood exterior (31%), interior (20%), railway (15%) (not applicable to the UAE) and industrial noise (20%).

In this study, Geographic Information System (GIS) technology was utilized to gain some understanding of the spatial distribution and content of a selection of the tweets collected through 2015 according to the source type, Figure 2. It is noted that the most annoying sources, such as music, are in densely populated districts within cities; there appears to be a link between highly populated areas and the frequency of complaints. It should be recognized from municipalities that the number of complaints will rise in these areas as the urban population expands in the UAE.

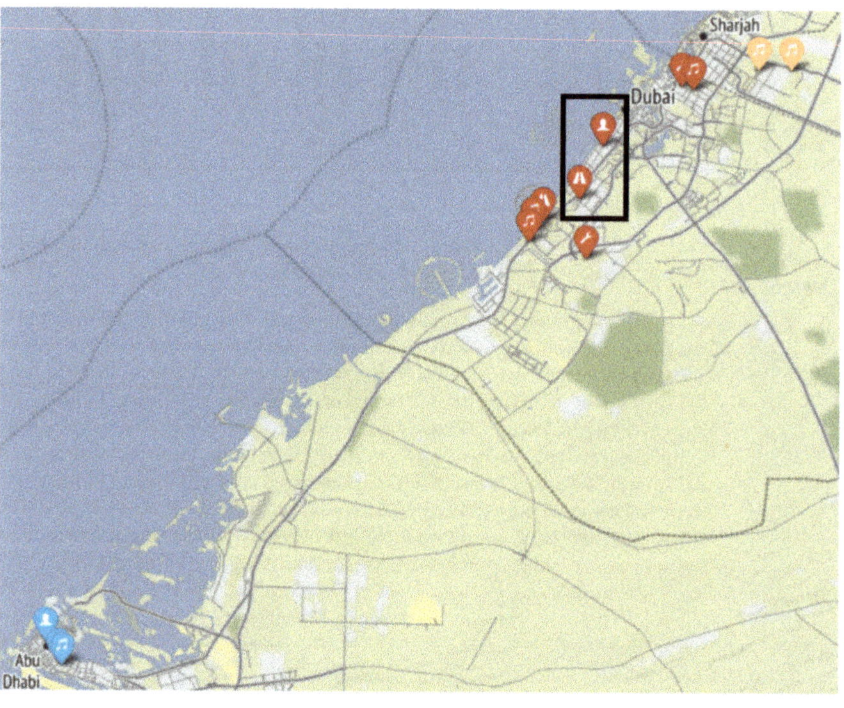

(**a**)

Figure 2. *Cont.*

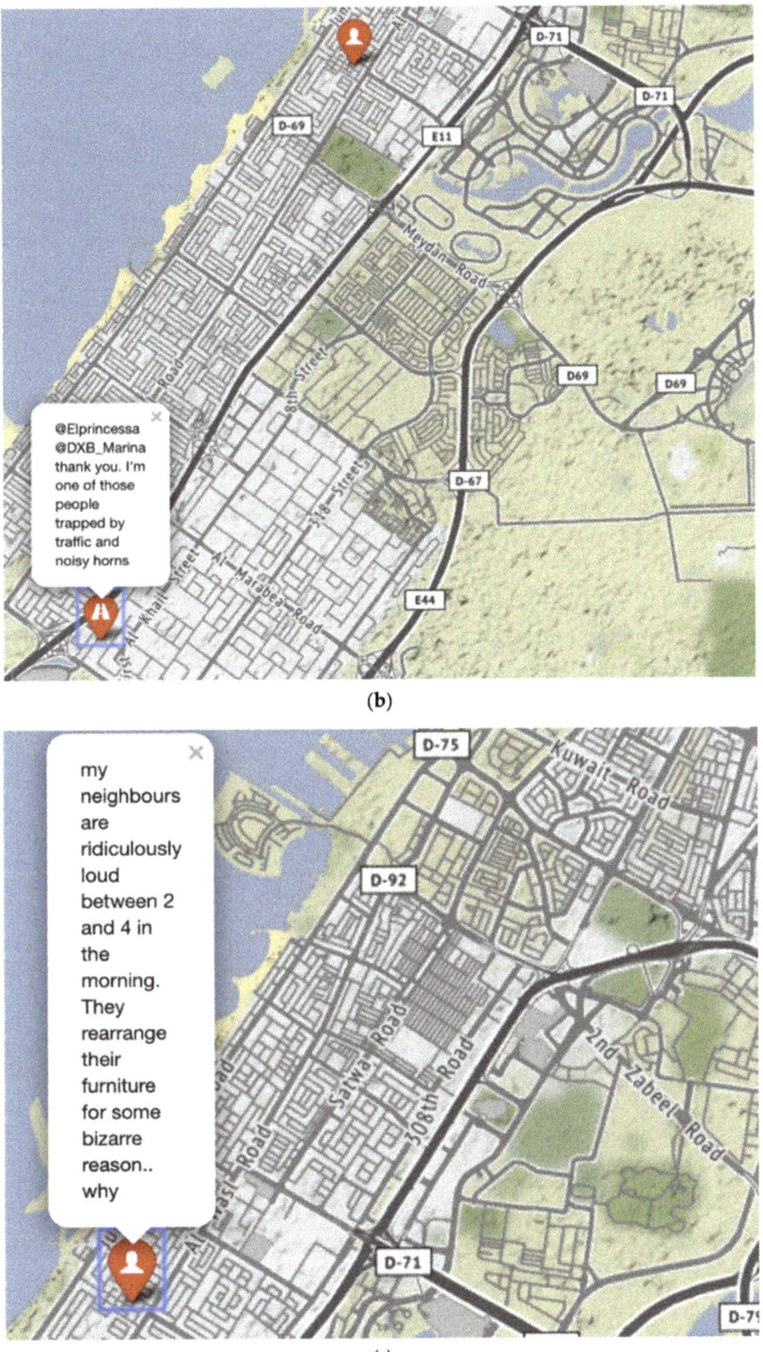

Figure 2. (**a**) Geolocation and characteristic of annoying noises, UAE Twitter 2015, blue = Abu Dhabi, red = Dubai, beige = Sharjah (**b**) reference to insert in (**a**): text "my neighbors are ridiculously loud between 2 and 4 in the morning. They rearrange their furniture for some bizarre reason", (**c**) text "*@Elprincessa @DXB_Marina thank you. I'm one of those people trapped by traffic and noise horns*". Legend icons: musical note = music; tool = construction; person silhouette = human; road = traffic; airplane = airplanes; building services = house.

4. Discussion and Conclusions

The use of big data has advantages over other forms of self-reporting in that it captures subjective noise complaints in a relatively naturalistic manner. Big data also has the potential to provide surveillance style reports based on larger datasets spanning multiple years. That said, big data provides a heuristic level of analysis that could form part of a larger, triangulated assessment plan providing cross-validation to objective noise measures and more traditional self-report measures. The representation of sound sources, which were obtained from "tweeting" in social media, including music and neighborhood noise, is affected by several biases since tweeting is an instantaneous reaction to a stimulus. Not all residents have access to social media or immediate access when annoyance occurs. Moreover, many noises are not "available" to be immediately tweeted due to the location of the noise and the presence of the person able to report their findings. This also has a bias on location finding since there could be a large error in the position of the original source. Nevertheless, without the onerous task of manually checking each tweet, it possible to train the query search to accept or decline genuine and accurate data points via machine learning or a knowledge base. Any form of knowledge base could include a larger lexicon than the non-exhaustive example we propose, which could include slang, for example. In the present study, the volume of our dataset resulted in a modest 38 hits, a severe limitation of the team's present access to the Twitter data set. In future studies with access to a much larger dataset and computer Random Access Memory (RAM) storage availability, perhaps Twitter API open-access data for replicability could be exploited and spanning several years, languages and countries the methods trialed in the present study could prove to be a valuable method for exploring noise pollution and efforts to reduce it.

The present concept study explored the utility of using social media data as a heuristic means of measuring and locating noise pollution trouble spots. This is not to suggest that council services should be employed immediately based on freshly tweeted alarms, but that "annoyance maps" could be created to capture any trends in certain residential districts, for example, which may be noteworthy. Based on the 2015 dataset extracted from the social media platform Twitter, noise annoyance times, locations and sources were identified. Public health statistics worldwide indicate that airport and traffic noise carries the most weight towards medical health problems but targeting and labeling a specific characteristic for noise, which causes the most "annoyance," is still an open problem. From the small sample extracted in this case study, the data suggest neighborhood or public-entertainment music, not traffic-noise, as the main culprit for "immediate" personal annoyance. Although this study concerns noise, which is an unwanted sound, it could be used in determining areas, which could benefit urban planners or researchers to shape a good "soundscape".

Author Contributions: Conceptualization, J.T. and A.P.; methodology, J.T., A.P. and A.A.; software, A.A.; formal analysis, A.A. and A.P.; investigation, J.T., A.P. and A.A.; resources, J.T. and A.A.; data curation, A.A.; writing—review and editing, J.T., A.P. and A.A.; All authors have read and agreed to the published version of the manuscript.

Funding: This research received no external funding.

Institutional Review Board Statement: Not Applicable

Informed Consent Statement: Not Applicable

Data Availability Statement: Data provided by Twitter through commercial agreement.

Conflicts of Interest: The authors declare no conflict of interest.

References

1. Freeman, C.; Francisco, L. *As Time Goes By: From the Industrial Revolutions to the Information Revolution*; Oxford University Press: New York, NY, USA, 2002.
2. Ginsberg, J.; Matthew, H.; Mohebbi, R.; Patel, S.; Brammer, L.; Smolinski, M.; Brilliant, L. Detecting influenza epidemics using search engine query data. *Nature* **2009**, *457*, 1012–1014. [CrossRef] [PubMed]

3. Eysenbach, G. Infodemiology: tracking flu-related searches on the web for syndromic surveillance. In *AMIA Annual Symposium Proceedings*; American Medical Informatics Association: Bethesda, MD, USA, 2006; pp. 244–248.
4. Al Shehhi, A.; Thomas, J.; Welsch, R.; Grey, I.; Aung, Z. Arabia Felix 2.0: a cross-linguistic Twitter analysis of happiness patterns in the United Arab Emirates. *J. Big Data* **2019**, *6*, 33. [CrossRef]
5. Dodds, P.; Harris, K.; Kloumann, I.; Bliss, C.; Danforth, C. Temporal Patterns of Happiness and Information in a Global Social Network: Hedonometrics and Twitter. *PLoS ONE* **2011**, *6*, e26752. [CrossRef] [PubMed]
6. Chao, Y.; Srinivasan, P. Life Satisfaction and Pursuit of Happiness on Twitter. *PLoS ONE* **2016**, *11*, e0150881.
7. Thomas, J.; Al Shehhi, A.; Grey, I. The sacred and the profane: social media and temporal patterns of religiosity in the United Arab Emirates. *J. Contemp. Relig.* **2019**, *34*, 489–508. [CrossRef]
8. Thomas, J.; Al-Shehhi, A.; Al-Ameri, M.; Grey, I. We tweet Arabic; I tweet English: self-concept, language and social media. *Heliyon* **2019**, *5*, e02087. [CrossRef] [PubMed]
9. Wenbo, W.; Chen, L.; Thirunarayan, K.; Sheth, A. Cursing in English on Twitter. In Proceedings of the 17th ACM conference on Computer supported cooperative work & social computing, Baltimore, MD, USA, 15–19 February 2014.
10. Aiello, L.; Schifanella, R.; Querica, D.; Aletta, F. Chatty Maps: constructing sound maps of urban areas from social media data. *R. Soc. Open Sci.* **2016**, *3*, 150690. [CrossRef] [PubMed]
11. Casco, L.; Clavel, C.; Asensio, C.; de Arcas, G. Beyond sound level monitoring: Exploitation of social media to gather citizens subjective response to noise. *Sci. Total. Environ.* **2019**, *658*, 69–79. [CrossRef] [PubMed]
12. Gasco, L.; Schifanella, R.; Aiello, L.; Quercia, D.; Asensio, C.; de Arcas, G. Social Media and Open Data to Quantify the Effects of Noise on Health. *Front. Sustain. Cities* **2020**, *2*, 41. [CrossRef]
13. WHO and JRC Report. Burden of Disease from Environmental Noise. 2013. Available online: www.euro.who.int/__data/assets/pdf_file/0008/136466/e94888.pdf (accessed on 1 February 2017).
14. Beutel, M.; Jünger, C.; Klein, E.; Wild, P.; Lackner, K. Noise Annoyance Is Associated with Depression and Anxiety in the General Population- The Contribution of Aircraft Noise. *PLOS ONE* **2016**, *11*, e0155357. [CrossRef] [PubMed]
15. Passchier-Vermeer, W.; Passchier, W. Noise Exposure and Public Health. *Environ. Health Perspect.* **2000**, *108* (Suppl. 1), 123–131.
16. Marianna, J.; Mariusz, W.; Konrad, L.; Grzegorz, K. Noise and environmental pollution from transport: decisive problems in developing ecologically efficient transport systems. *J. Vibroeng.* **2017**, *19*, 5639–5655.
17. Öhrstrom, E.; Barregard, L.; Andersson, E.; Skanberg, A.; Svensson, H.; Angerheim, P. Annoyance due to single and combined sound exposure from railway and road traffic. *J. Acoust. Soc. Am.* **2007**, *122*, 2642–2652. [CrossRef] [PubMed]
18. Selander, J.; Nilsson, M.; Bluhm, G.; Rosenlund, M.; Lindqvist, M.; Nise, G.; Pershagen, G. Long-term exposure to road traffic noise and myocardial infarction. *Epidemiology* **2009**, *20*, 272–279. [CrossRef] [PubMed]
19. Sörensen, M.; Hvidberg, M.; Hoffmann, B.; Andersen, Z.; Nordsborg, R.; Lillelund, K.; Jakobsen, J.; Tjonneland, A.; Overvad, K.; Raaschou-Nielsen, O. Exposure to road traffic and railway noise and associations with blood pressure and self-reported hypertension: a cohort study. *Environ. Health* **2011**, *10*, 92. [CrossRef] [PubMed]
20. Theebe, M. Planes, trains, and automobiles: the impact of traffic noise on house prices. *J. Real Estate Finance Econ.* **2004**, *28*, 209–234. [CrossRef]
21. Noise in Europe 2014. EEA Report 10. 2014. Available online: www.eea.europa.eu/publications/noise-in-europe-2014 (accessed on 20 June 2018).
22. Environmental Protection Act (UK). 2004. Chapter 43. Available online: https://www.ecolex.org/details/legislation/environmental-protection-act-1990-chapter-43-lex-faoc005515/ (accessed on 1 February 2021).

International Journal of
Environmental Research and Public Health

Review

Airborne Sound Power Levels and Spectra of Noise Sources in Port Areas

Samuele Schiavoni [1], Francesco D'Alessandro [2], Davide Borelli [3,*], Luca Fredianelli [4,*], Tomaso Gaggero [5], Corrado Schenone [3] and Giorgio Baldinelli [6]

1. Metexis s.r.l., Via Bartolomeo Grazioli 91, 06132 Perugia, Italy
2. Italian National Research Council—Institute on Atmospheric Pollution at Italian Ministry of Ecological Transition, Via Cristoforo Colombo 44, 00147 Rome, Italy
3. Department of Mechanical, Energy, Management and Transport Engineering, Division of Thermal Energy and Environmental Conditioning, University of Genoa, Via all'Opera Pia 15/A, 16145 Genoa, Italy
4. Institute of Chemical and Physical Processes of National Research Council, Via G. Moruzzi 1, 56124 Pisa, Italy
5. Department of Telecommunications, Electrical and Electronics Engineering and Naval Architecture, University of Genoa, Via Montallegro 1, 16145 Genoa, Italy
6. CIRIAF—Inter University Research Centre for Environment and Pollution "Mauro Felli", Via G. Duranti 67, 06125 Perugia, Italy
* Correspondence: davide.borelli@unige.it (D.B.); luca.fredianelli@ipcf.cnr.it (L.F.); Tel.: +39-01-03-352-572 (D.B.)

Citation: Schiavoni, S.; D'Alessandro, F.; Borelli, D.; Fredianelli, L.; Gaggero, T.; Schenone, C.; Baldinelli, G. Airborne Sound Power Levels and Spectra of Noise Sources in Port Areas. *Int. J. Environ. Res. Public Health* **2022**, *19*, 10996. https://doi.org/10.3390/ijerph191710996

Academic Editor: Paul B. Tchounwou

Received: 4 August 2022
Accepted: 30 August 2022
Published: 2 September 2022

Abstract: Airborne port noise has historically suffered from a lack of regulatory assessment compared to other transport infrastructures. This has led to several complaints from citizens living in the urban areas surrounding ports, which is a very common situation, especially in countries facing the Mediterranean sea. Only in relatively recent years has an effort been made to improve this situation, which has resulted in a call for and financing of numerous international cooperation research projects, within the framework of programs such as EU FP7, H2020, ENPI-CBC MED, LIFE, and INTERREG. These projects dealt with issues and aspects of port noise, which is an intrinsically tangled problem, since several authorities and companies operate within the borders of ports, and several different noise sources are present at the same time. In addition, ship classification societies have recently recognized the problem and nowadays are developing procedures and voluntary notations to assess the airborne noise emission from marine vessels. The present work summarizes the recent results of research regarding port noise sources in order to provide a comprehensive database of sources that can be easily used, for example, as an input to the noise mapping phase, and can subsequently prevent citizens' exposure to noise.

Keywords: port noise; noise sources; noise mapping; noise mitigations; noise modeling; ship noise; sustainable management; noise exposure prevention; noise measurements; research projects

1. Introduction

Nowadays, the world economy is globally interconnected, which means that every year, larger amounts of goods need to be transported between countries and continents in a safe, efficient and competitive way. Since the second half of the twentieth century, when intermodal container shipping was invented in the U.S., this method of freight transportation has gained more and more importance, leading to the creation of large ports all over the world, with Asia leading the chart with the largest container ports. Over 80% by volume of the international goods traded are carried by sea, and the percentage is even higher for most developing countries. The COVID-19 pandemic affected maritime transport but the effects were lower than expected [1], even if, in February 2022, more than eleven percent of the global container ship capacity was unused, and the average shipping delay has also increased due to port congestion [2].

In addition to the maritime traffic generated by container shipping, passenger ships, and especially the leisure cruise industry, have experienced an annual passenger growth rate of 6.6% from 1990 to 2019. In this case, the effect of the pandemic was definitely greater, with almost one year of a complete stop on passenger cruises. On the other hand, the pandemic also accelerated the retirement of many ships leading to more modern and environmentally friendly fleets [3].

In light of the above, ports are crucial elements for the global market, but they may generate severe negative impacts, mostly related to the environment, land use and traffic congestion. The main negative environmental impacts are due to the emission of noise, odors and volatile organic substances, and to the pollution of water and soil by oil chemicals, hull paint and other hazardous materials [4]. Furthermore, most of these negative impacts are localized, taking place close to the port (in terms of noise and dust) and in the urban area (for air emissions, water quality, congestion and land use) [5]. In several cases, ports are located in close proximity to urbanized areas and they may even be bounded by or include environmentally protected areas [6]. Thus, it is evident how global needs and benefits produce local negative impacts.

From the noise point of view, ports are complex infrastructures if compared to other transport infrastructures (roads, railways, airports) and logistic nodes. The possible sources of noise that can be found in a port range from ships in transit to stationary ships, generators, maneuvering equipment, cranes, machinery and ventilation systems, but also moving vehicles and trains. Fredianelli et al. [7] made a comprehensive list of noise sources that can be found in a port area, which is reported in Section 2 of the present paper.

The great number of different sources of noise are dynamically distributed in space and time in a relatively large area, which is usually characterized by unsteady behavior or tonal components. The result is a sound environment with an extraordinarily variable temporal and spectral structure, where single sound sources are difficult to isolate. Furthermore, many of them are characterized by prominent low frequency components, between 20 Hz and 200 Hz, which can travel long distances with limited attenuation and are hardly insulated from building walls [8].

For all of the above-mentioned reasons, ports are complex environments for professionals to deal with, also because specific noise-related rules and regulations for ports are often missing, both at the international and national level. For instance, the European Directive 2002/49/EC on the Assessment and Management of Environmental Noise (END) is focused on two main categories of noise sources: transport infrastructures and industrial sites. Even if the END specifies that industrial (port) areas near large agglomerations must be included in noise maps, it gives no specific indications on how to draw their noise maps. As a consequence, noise nuisance in port areas is usually addressed by considering the ports as a single noise source, similar to an industrial area, with evident underestimation of the issues. With the absence of a uniform approach or guidelines, when problems arise regarding citizens affected by port noise [9], it is generally addressed at a local level by following different approaches that only tend to respond to complaints. In Italy, as a matter of example, even if specific decrees that regulate the noise produced by roads, railways, airports and industries exist, a national decree for the regulation of noise generated by port activities is still missing, although it is required by national law.

Ports, as noise sources, and industrial noise in general, are neglected by the World Health Organization too, with its 2019 environmental noise guidelines for the European region [10], providing policy makers with recommendations for protecting human health from exposure to environmental noise originating from transportation (road traffic, railway and aircraft), wind turbine noise and leisure noise. As was observed by Bernardini et al. [11], unlike the broad literature on transportation noise and industrial noise and the variety of mitigation measures for those sources, the scientific community paid very little attention to analyzing and tackling the noise produced by ports and the effects of the exposure on the surrounding population. Furthermore, the majority of scientific research regarding the noise generated from maritime transport is focused on studying onboard vessel noise, its interference with animal life or oceanic ambient noise. In this field of research, in fact, underwater noise is investigated more than airborne noise.

Evidently, planning or managing noise in port areas can be an overwhelming task for even the most skilled professional, especially when it comes to simulating noise propagation and evaluating noise levels at the receivers' location. The main problem is nearly always in defining the noise sources that are located inside the port area, even if the definition of the port boundary can be a problem on its own [12], in addition to where and for how long the sources are operating, and most importantly, their characterization in terms of directivity and sound power.

The present study reviews the current scientific literature, technical report and other databases with the aim of collecting the sound power levels of the different noise sources acting in port areas. The comprehensive list that is created would be very useful for professionals and technicians as input data for simulation and the mapping phase needed in the noise management toward citizens' health care. Moreover, the work also acts as a further starting point for driving future work into a harmonized approach of study regarding port noise.

The following sections of this paper are structured as follows: Section 2 summarizes the recently published guidelines for noise mapping of port areas, Section 3 collects the noise emission data of major port sources, divided according to the most used typology: transtainer, reach stacker, straddle carrier, gantry cranes, reefer, moored ships, Ro-Ro and Ro-Pax ramps, forklifts, and seagoing ships. Section 4 provides the conclusions to the work.

2. Available Source Characterization Guidelines for Noise Mapping of Port Areas

A joint product of the Interreg European projects REPORT, MON ACUMEN, DECIBEL and RUMBLE was the development of source characterization guidelines for noise mapping of port areas. As described by Fredianelli, et al. [13], the aim of the work was to present specific measurement procedures for the assessment of noise emissions of the many sources acting in ports, according to the five macro-categories and further sub-categories previously proposed in another work [7]. The procedure was set in order to provide technicians and stakeholders with a unified methodology that allows the retrieval of the inputs for noise mapping software. The guideline expectations were to boost sector studies and start providing a common approach to acoustic mapping of ports that will allow a proper comparison of population exposure levels in the future.

The categories for which a specific procedure was reported are:

- Road:
    a. Internal traffic;
    b. Port-related external traffic;
    c. External traffic not generated by the port.
- Railways:
    a. Internal traffic;
    b. Port-related external traffic;
    c. External traffic not generated by the port.
- Ships:
    a. Sailing at a reduced speed approaching the quay;
    b. Moored in stationary conditions;
    c. Mooring operations;
    d. Moored during loading/unloading operations (without auxiliary machinery).
- Port and industrial:
    a. Fixed sources;
    b. Mobile sources;
    c. Area sources.

In fact, all the sources falling into the previously mentioned categories have peculiarities that led to a different noise emission with respect to the others, which obviously translates into the need for a specific measurement procedure. Only minor adaptations to port environment scenarios were proposed for roads and railways, which are well characterized sources with CNOSSOS-EU as a proper model [14]. The modifications considered the high percentage of heavy and freight vehicles with respect to passenger vehicles in the port infrastructure, with the annexed average reduction in speed.

All of the choices made in defining the measurement procedures for ships, port and industrial sources were made assuming the lower interference with port and ship operations and the impossibility to measure onboard. In addition, the measurement procedures do not need the collaboration from both ships or terminal owners, in addition to switching the machinery off/on on request to reduce the background noise. Furthermore, the use of cranes or cherry pickers to reach a higher measurement position was discarded for the sake of simplicity and economy. A simplification made in the work was considering ships as the emitting noise in a symmetric way with respect to its vertical longitudinal symmetry plane, even if this assumption is not true for some vessels. This allows technicians to perform characterization measurements on only one side and then avoiding renting boats to reach the side facing the sea of the moored ship.

Port and industrial categories include the same machinery and vehicles used in port activities or for the industries in the port area, but a different classification is needed for better identifying the legal responsibility of limit exceedances. As both categories are very wide, a further subdivision was carried out in the moving or fixed source, or even the area source when details are not important. Pumps, generators, ventilators, air conditioners, machinery of any type, fixed cranes, conveyor belts and refrigerated containers are some of the fixed sources that can be found in a port environment. Instead, straddle carriers, frontlifts, contstackers, forklifts, transtainers, cranes, dock tractors and other cargo handling units are the mobile sources. The measurement procedure for a mobile source is different according to its operation phase (transit, handling, loading/unloading operations), which should be all properly characterized.

All of the noise measurements must be performed with a class I [15] sound level meter. The sound power level (L_W) of the investigated source is calculated from the sound pressure levels (L_p) obtained using Equation 1, possibly separated in third octave bands.

$$L_W = L_p + 10 \cdot \log[Q/(4\pi r^2)], \tag{1}$$

where Q is the directivity factor and r is the distance to sound source.

The resulting sound power level L_W is the information needed by the models to simulate the noise in port areas.

For the purpose of the present work, however, the guidelines represent the first attempt to drive the collection of sound emitted by port sources with the intent of stimulating the scientific community to create an accessible and comprehensive database.

3. Collection of Noise Emission Data of Major Port Sources

The following reports a selection of the most solid and consistent data concerning sound emitted by sources located inside port areas, retrieved from the scientific and technical literature as well as from the reports of the European projects cited in the previous paragraphs; data are reported as sound power level L_W for each source and expressed in dB or dB(A) for point sources, in dB/m or dB(A)/m for linear sources and in dB/m^2 or dB(A)/m^2 for area sources.

The retrieved information regarding the sound power levels of noise sources operating in port areas are mainly gathered from:

- The outcomes of the REPORT project [16] for transtainer, reach stacker, reefer and gantry cranes;
- The paper "Noise evaluation of sound sources related to port activities" [17] for Ro-Ro and Ro-Pax vessels, rubber-tired gantries, straddle carriers, reach stackers, reefers and the ramp noise, realized within the activity of the EU funded EFFORTS project [18];
- The paper "Container Terminals and Noise" for container ships, straddle carriers and tractors [19];

- The report "Technical noise investigations at Hamburg City cruise terminals" of the INTERREG Green Cruise Port for moored cruise ships, reefers and forklifts [20];
- The outcomes of the NEPTUNES project for ships and other sources [21];
- The paper "Noise emission Ro-Ro terminals" for Ro-Ro moored ships [22];
- The paper "Airborne noise emissions from ships: Experimental characterization of the source and propagation over land" for container ships [23];
- The paper "Evaluation and control of cruise ships noise in urban areas" for cruise ships [24];
- The report "Noise from ships in ports" for moored ships and reefers [25];
- The report of the Lloyd's Register regarding how the noise emissions of a moored ship have to be modelled [26];
- The report "Assessment of the acoustic benefit of the power supply to ships moored in ports (cold ironing)" [27] and the related paper presented at the Euronoise 2018 Conference [28];
- The paper "Pass-by Characterization of Noise Emitted by Different Categories of Seagoing Ships in Ports" [29];
- The outcomes of the FP7 SILENV project for the moored ships ([9,30–33]);
- ISPRA (Italian National Institute of Environmental Protection and Research) data based on the FP7 SILENV project [34,35].

A summary of the standards used in the acoustic measurements carried out to define the sound power level of the port noise sources described in the documents above can be found in Appendix A.

3.1. Transtainer

The REPORT project [16] gives the one-third octave band's sound power level spectra of a transtainer on standby, in movement with the alarm signal horn functioning and in full activity. The standby emission is represented by a point source; the moving transtainer with the alarm signal being modelled as a linear source. The activity of the transtainer is made by all the movements it makes to take, move and place containers in port areas. Data collected from noise measurement in the REPORT project evidence that transtainer emissions on the railway side are different from the ones on the square side. Figure 1 reports the sound power emission spectra of a transtainer, as calculated in the REPORT project:

- TR-A-CA-RW-WH considering an equivalent area source, device performing a complete activity, measurement focused on the railway side;
- TR-A-CA-SQ-WH, same as the previous item, but focused on the square side;
- TR-L-MV-RW-WH considering a linear noise source representing the movement of the transtainer on the rail, measurement focused on the railway side;
- TR-L-MV-SQ-WH considering a linear noise source representing the movement of the transtainer on the rail, measurement focused on the square side;
- TR-P-STBY-WH considering a point noise source representing the transtainer on standby. The point noise source can be used to model the emission of the whole device.

Another characterization of this noise source was performed within the activities of the EFFORTS project [17,18]. The noise emission characteristics of the device was defined by three different point sources representing the power unit, the exhaust pipe (20 m above the ground) and the alarm signal. Figure 2 reports the following data:

- TR-P-LIFT-AL is the sound power level of a point noise source representing the noise emission of the alarm signal when the transtainer is performing an operation of lifting or picking up containers;
- TR-P-LIFT-FU is the sound power level of a point noise source representing the noise emission of the funnel when the transtainer is performing an operation of lifting or picking up containers;
- TR-P-LIFT-PU is the sound power level of a point noise source representing the noise emission of the power unit when the transtainer is performing an operation of lifting or picking up containers;

- TR-P-STBY-FU is the sound power level of a point noise source representing the noise emission of the funnel when the transtainer is on standby (idling);
- TR-P-STBY-PU is the sound power level of a point noise source representing the noise emission of the power unit when the transtainer is on standby (idling).

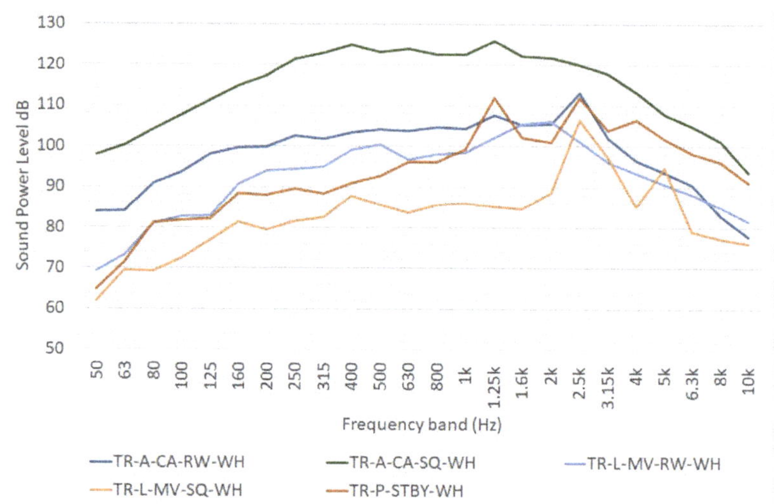

Figure 1. One-third octave sound power level spectra of a transtainer, as reported in [16]. L_W is expressed in dB for point sources, in dB/m for linear sources and in dB/m^2 for area sources.

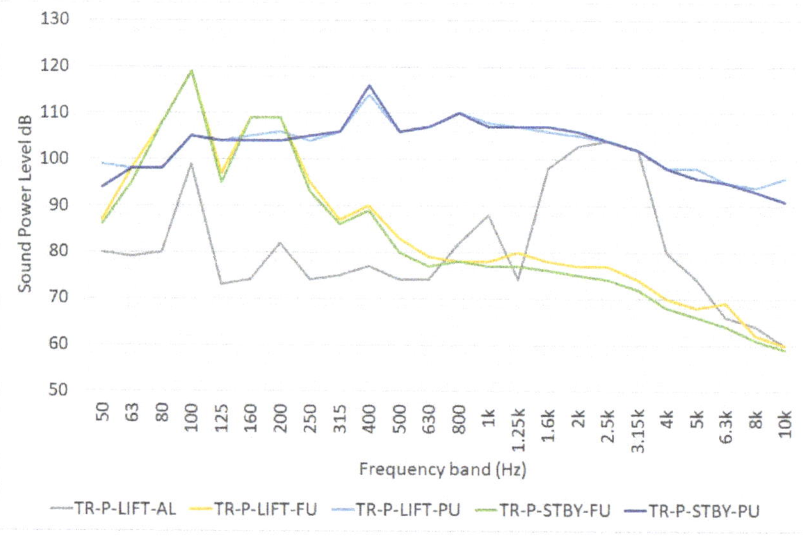

Figure 2. One-third octave sound power level spectra of a transtainer, as reported in [17,18].

Moreover, the noise emission of the equipment was obtained through measurement under idling and lifting conditions.

Sound power data from the REPORT Project [16] represent the noise source as a single source, so it can be handled more easily than those of the EFFORT Project [18]. Nevertheless, it is worth nothing that noise measurements used for calculating the sound power level in [18] were carried out in compliance with ISO 3744. There is no information about the standard of the noise measurement carried out in [16].

3.2. Reach Stacker

The REPORT [16] and the EFFORTS projects [17,18] estimated the noise emitted by reach stackers. Figure 3 reports the following data:

- RS-A-CA-RW-WH and RS-A-CA-SQ-WH are the sound power levels of two equivalent areal noise sources representing the noise emission of two different reach stackers performing a complete activity, as estimated in [16];
- RS-P-LIFT-WH and RS-P-LIFT-WH (2) are the sound power levels of point noise sources representing the noise emission of the reach stacker when it is performing an operation of lifting or picking up containers, respectively, estimated by [18,36]. The latter was derived from the global sound power level ($L_{w,sum}$), considering a pink noise source;
- RS-P-PB-WH is the sound power level of a point noise source representing the noise emission of the device pass-by [18];
- RS-P-SG-WH is the sound power level of a point noise source representing the noise emission of the device performing an operation of setting containers to the ground [18];
- RS-P-STBY-WH is the sound power level of a point noise source representing the noise emission of the device in standby mode (idling) [16];

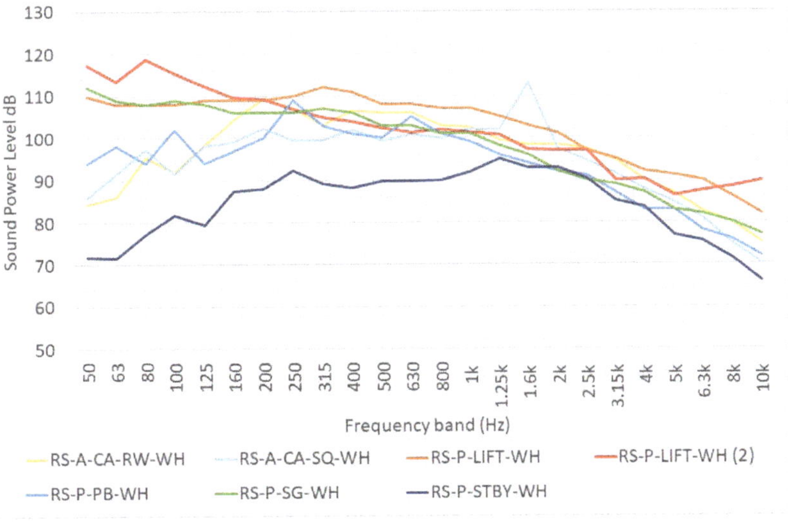

Figure 3. One-third octave sound power level spectra of a reach stacker, as reported in [16–18,36]. L_W is expressed in dB for point sources, in dB/m for linear sources and in dB/m² for area sources. RS-A-CA-RW-WH, RS-A-CA-SQ-WH and RS-P-STBY-WH data are taken from [16], RS-P-LIFT-WH, RS-P-PB-WH and RS-P-SG-WH from [18], RS-P-LIFT-WH (2) from [36].

3.3. Straddle Carrier

The noise emission of a straddle carrier has been determined using the pass-by method by [17] within the activity of the EFFORTS project: the sound power level of the equivalent point noise source SC-P-PB-WA is reported in Figure 4 and in Table 1. The overall sound power level was reported in the deliverable 2.4.3 of the project [18]. Another remarkable study investigating noise emission from straddle carriers were carried out by Witte [19].

This study does not provide spectral information, but only reports the overall sound power level in dB(A), without specifying how these data were calculated (Table 1).

Table 1. Sound power level of straddle carriers as reported in [18,19].

Straddle Carrier Activity	L_W (dB)	L_W (dB(A))
Pass-by [18]	119 ± 2	115 ± 2
Normal activity, power unit close to the ground [19]	/	108
Normal activity, power unit located at the top [19]	/	104

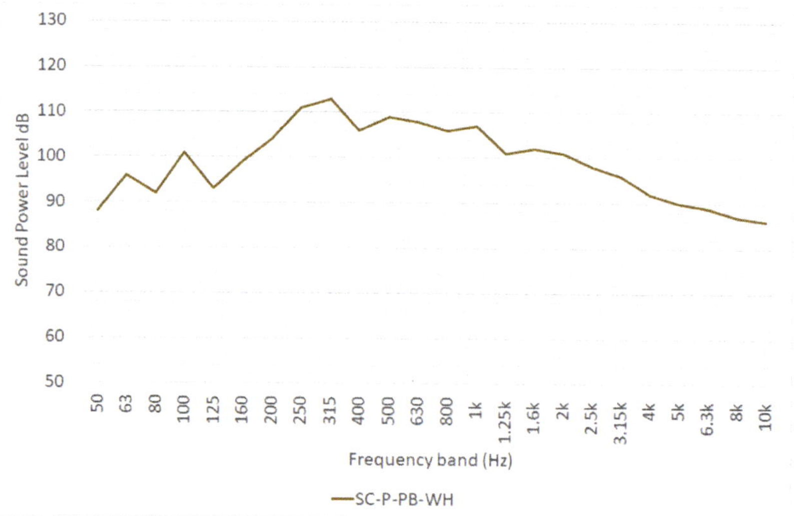

Figure 4. One-third octave sound power level spectrum of a straddle carrier pass-by, as reported in [17]. L_W is expressed in dB/m for linear sources.

3.4. Gantry Cranes

The REPORT [16] and the EFFORTS projects [17,18] estimated the sound power level spectra of gantry cranes. Figure 5 shows the following data:

- GC-A-CA-WH is the sound power level of two equivalent areal noise sources representing noise emission due to the complete activity of a gantry cranes, as estimated in [16];
- GA-P-CA-WH and GA-P-LIFT-WH are the sound power level spectra of point noise sources representing the noise emission of the complete activity and of the lifting operation alone, respectively. These data were estimated in [17,18].

These data were obtained by the analysis of noise measurements. Witte provided a rough estimation of sound power level equal to 100 dB(A) for gantry cranes in [19], without providing further information relating to how the data were obtained (noise measurements, databases, etc.).

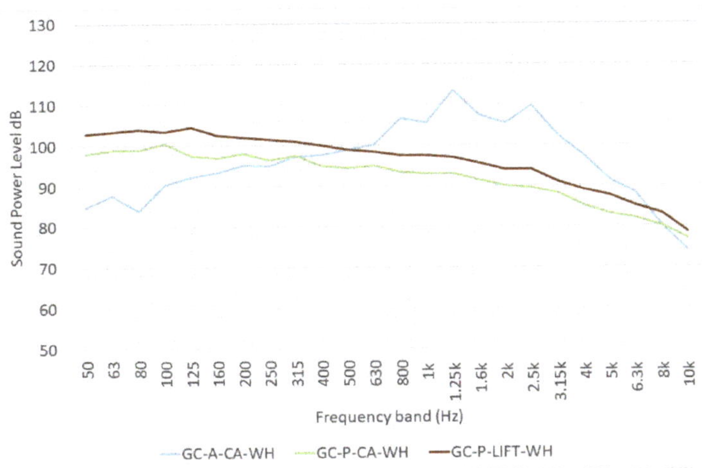

Figure 5. One-third octave sound power level spectra of gantry cranes, as reported in [16–18]. L_W is expressed in dB for point sources, in dB/m for linear sources and in dB/m^2 for area sources. GA-P-CA-WH and GA-P-LIFT-WH data are taken from [17,18], GC-A-CA-WH from [16].

3.5. Reefer

Several data are available relating to the assessment of the noise emission of reefers, i.e., refrigerated containers. The characterization of this kind of equipment is easier in comparison to other port noise sources since it is a regular container with an HVAC unit devoted to maintaining an adequate temperature inside.

Noise measurements were used to characterized the one-third octave acoustic emission of reefers in both the REPORT project [16] and the EFFORTS project [17,18] (Figure 6); the figure also contains octave band data taken from a 2010 report of the Danish Ministry of Environment [25].

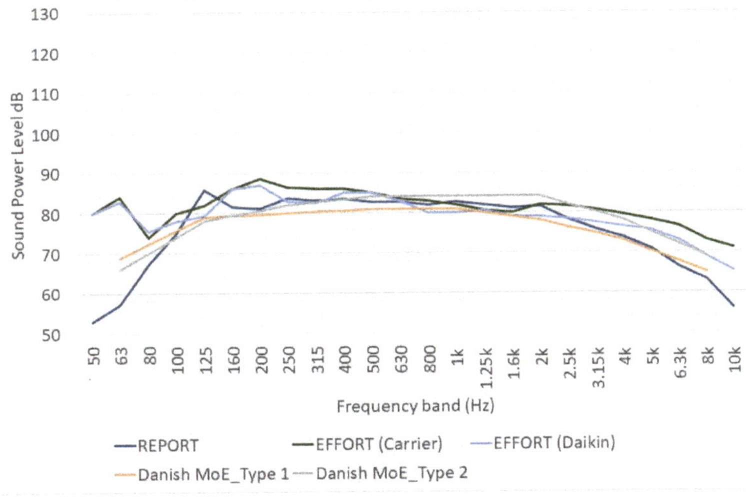

Figure 6. Sound power level spectra of reefers, as defined by REPORT Project [16], EFFORT Project [18] and a report of the Danish Ministry of Environment [25]. L_W is expressed in dB for point sources, in dB/m for linear sources and in dB/m^2 for area sources.

Other noise emission assessments of reefers based on measurements have been carried out by the NEPTUNES project [21] and by the Interreg-funded project Green Cruise Port [20] (see Table 2).

Table 2. Sound power level of reefers, as reported in [20,21].

Source	Working Condition	L_W (dB(A))
NEPTUNES Project [21], mitigation of noise from ships at berth	Normal activity	91–93
GREEN CRUISE PORT [20]	Normal activity	99

The differences observed in terms of noise emission are probably caused by the different models of the cooling units installed in each device. It is worth nothing that only data in REPORT project give additional information regarding how the noise measurements were carried out.

3.6. Moored Ships

The best practice guide "Mitigation of Noise from Ships at Berth" of the NEPTUNES project [21] suggests that ships may be divided into six different classes:

- Container ships;
- Cruise ships;
- Tankers;
- Ro-Ro and Ro-Pax;
- Bulk Carriers;
- General cargo/ service ships.

Their noise emissions are caused by:

- The funnel outlet(s) of the auxiliary engine(s), all ship types;
- The opening of the engine room ventilation inlet(s) and outlet(s), all ship types;
- The opening of the cargo hold ventilation and air conditioning inlet(s) and outlet(s), all ship types;
- The opening of the ventilation and air-conditioning of passenger rooms (cruise ship and Ro-Pax);
- Further relevant ventilation openings;
- Pumps on deck (tankers).

The project gives some indicative values of the noise emission of a moored ship; they are reported in Table 3. These data cannot be used for noise assessment studies because they oversimplify the complexity of the acoustic emission of a moored ship. However, they can give an indication regarding the impact of the noise emission of a ship at berth without a cold ironing solution.

Table 3. Approximate sound power level of some port noise sources given by [21].

Source	L_W (dB(A))
Container ship	100–115
Ro-Ro ship	100–114
	(1)

A more detailed study regarding the assessment of the noise emission of a ship at berth was carried out by the Danish Ministry of the Environment [25]. The study reports the sound power level of some types of diesel generators without silencers (Table 4) and ventilation fans (Table 5) used in vessels, collected from the producers of these components.

The data evidences that the anti-noise treatment of the engine is crucial to reducing the noise impact of a moored ship.

Table 4. Unattenuated one-third octave sound power level in dB(A) of the different diesel engine exhausts [25].

Producer	Type	One-Third Octave Frequency Bands (dB(A))										Total (dB(A))
		16 Hz	31.5 Hz	63 Hz	125 Hz	250 Hz	500 Hz	1000 Hz	2000 Hz	4000 Hz	8000 Hz	
	L32/40	80	115	130	135	129	133	135	135	133	130	142
	V32/40	82	111	126	133	129	133	135	135	133	130	142
MAN B and W	L48/60B	88	119	124	126	129	133	135	135	133	130	141
	V48/60B	84	111	124	126	129	133	135	135	133	130	141
	L58/64	80	115	130	135	129	133	135	135	133	130	142
	W26	-	122	132	135	131	125	124	118	112	102	138
Wärtsilä	W32	-	107	115	127	130	129	127	121	109	-	135
	W38	-	101	119	122	127	131	134	129	126	118	138

Table 5. One-third octave sound power level in dB(A) of different ventilation fans used in vessels without anti-noise measures [25].

Fan Function	Volume Flow (m³/h)	One-Third Octave Frequency Bands (dB(A))								Total (dB(A))
		63 Hz	125 Hz	250 Hz	500 Hz	1000 Hz	2000 Hz	4000 Hz	8000 Hz	
	120,000	73	93	98	105	105	102	98	91	110
	70,000	68	84	100	104	106	103	99	93	110
	50,000	66	82	98	101	103	101	97	90	108
Engine room fans	33,000	64	79	96	99	101	99	94	88	106
	15,000	51	67	80	95	96	96	92	86	101
	12,000	52	68	81	96	96	96	93	87	102
	1000	39	55	73	78	83	83	80	74	88
	95,000	75	93	97	100	100	97	91	83	105
Hold ventilation	85,000	69	89	94	101	101	98	94	87	106
	73,000	67	83	99	102	104	102	97	91	109

It is worth noting that a good approximation of the sound power level of a fan is given by Equation (2), reported in [37]:

$$L_W = L_W{}^* + 10 \cdot q_v + 20 \cdot \Delta p_v,\qquad(2)$$

where:

- $L_W{}^*$ can be assumed to be 25–30 dB for radial ventilators and 25–35 dB for axial ventilators;
- q_v is the volume flow in m³/h;
- Δp_v is the fan total pressure difference in Pa.

The noise emission of a Ro-Ro vessel at berth was also studied in the EFFORTS project ([17,18]); the sound power level spectra were obtained from noise measurements performed in the ports of Turku and Dublin (Figure 7). The overall sound power levels of these sources provided by the EFFORT project have been compared with similar data reported in a 2018 technical report delivered by Tecnalia [27] (Table 6). Both studies considered a Ro-Ro ship at berth; it is worth noting that the noise source "Auxiliary engine" in [27] likely groups together some of the sources considered in [17].

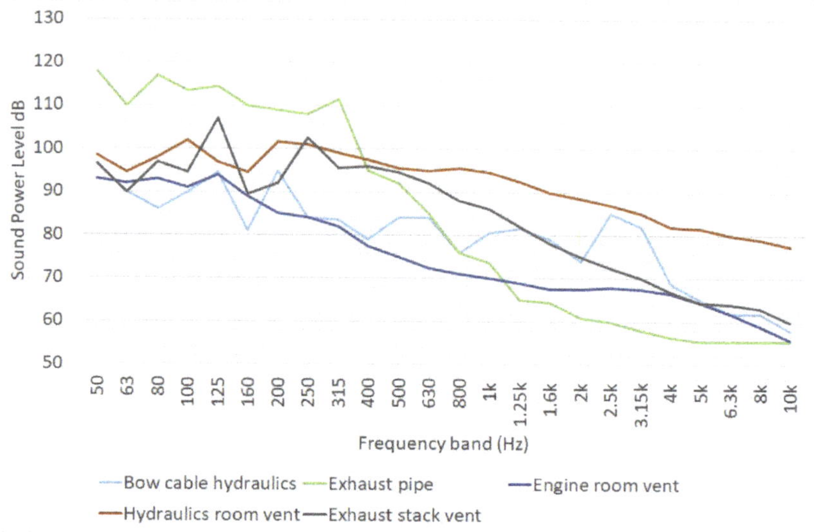

Figure 7. One-third octave sound power spectra of Ro-Ro vessel noise sources, as reported in [17]. L_W is expressed in dB for point sources, in dB/m for linear sources and in dB/m^2 for area sources.

Table 6. Comparison between data of EFFORT Project [18] and Tecnalia Report [27] related to the noise emission of some noise sources of Ro-Ro ships.

Ship Source	L_W (dB)	L_W (dB(A))
Engine room ventilation [18]	102	86
Hydraulic room ventilation [18]	110	104
Bow cable hydraulics [18]	103	93
Auxiliary engine exhaust pipe [18]	124	106
Exhaust stack ventilation [18]	112	100
Auxiliary engine [27]	/	107
Ventilation unit [27]	/	109

An approximate and quick assessment of the noise emission of container ships was carried out by Witte [19]: the relationship (Equation (3)) between the deadweight tonnage (DWT) and the A-weighted sound power level of a container ship can be expressed as follows:

$$L_{W,A} = 55.4 + 12.2 \cdot DWT \tag{3}$$

The equation was obtained by noise measurements on 65 ships. The use of this data to characterize the noise emission of a container ship has some drawbacks:

- The container ship is modelled through a single point noise source. This approach may lead to relevant errors in the assessment of noise impacts, in particular for receivers located close to the docks;
- The author does not provide detailed information about how the noise measurements were carried out and processed.

An accurate characterization of a container ship has been performed in [23], considering a detailed digital model of the vessel and the noise emission spectra of each source. The emission spectra were obtained by tailored measurements. In order to validate the noise model, horizontal and vertical grids of noise measurement were performed. The outcomes proved that "a limited number of dedicated onsite measurements together with adaptations of the code to the specific case allowed us to obtain an effective model for the ship".

The report of the Interreg-funded project Green Cruise Port [20] provides much information. Noise emission data of cruise ships were obtained from noise measurements considering separate exhaust gas outlets (Figure 8) and the ventilation openings of three vessels: AIDAsol (Length equal to 253/Width 38 m), AIDAprima (300/48 m) and Mein Schiff 3 (294/39 m).

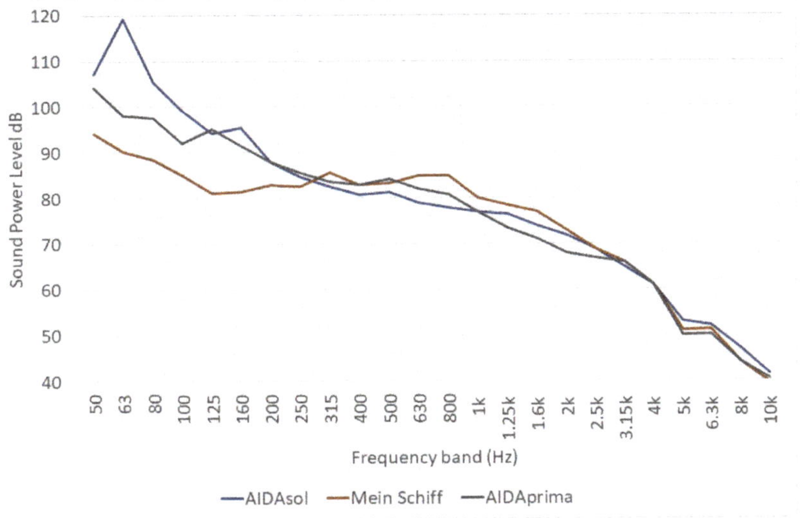

Figure 8. One-third octave sound power spectra of cruise ship funnel of three vessels, as reported in [20]. L_W is expressed in dB for point sources, in dB/m for linear sources and in dB/m^2 for area sources.

Concerning the exhaust gas outlet group, the three vessels have similar noise emission spectra for frequencies higher than 200 Hz. Under this threshold, the noise emission of the AIDAsol (the smaller one) is more relevant than for the other two ships. The equivalent sound pressure level $L_{w,A}$ is of 102 dB(A) for AIDAsol, 100 dB(A) for AIDAprima and 98 dB(A) for Mein Schiff 3. Noise measurements for the characterization of exhaust gas outlets were carried out in compliance with DIN 45635-47.

Concerning ventilation openings, noise measurements carried out on the three vessels evidenced that they may have a tonal (peak at 100 Hz) or a broadband character. The elaboration of the noise measurements performed in the Green Cruise Port project evidenced that the noise emission from the two side of the same cruise ship can be substantially different (Figure 9). Noise measurements carried out for the characterization of ventilation openings were carried out in compliance with DIN EN ISO 3746 [38].

In 2009, Witte carried out a noise measurement campaign to define a relationship between the loading capacity and noise emission of Ro-Ro ships, such as the one defined for container ships. However, in this case, the author did not find any relationship between the two parameters, as reported in [22].

A 2013 paper by Di Bella and Remigi reports the one-third sound power spectra of several typologies of cruise ships; they were evaluated as single-point noise sources and their sound power level was estimated from the elaboration of in-field measurements performed in compliance with several ISO standards. The spectra of these sources are reported in Figure 10 [24]. Nevertheless, the outcomes of some studies evidence that the characterization of a moored ship as a single-point noise source seems to be an excessive approximation ([33,39]).

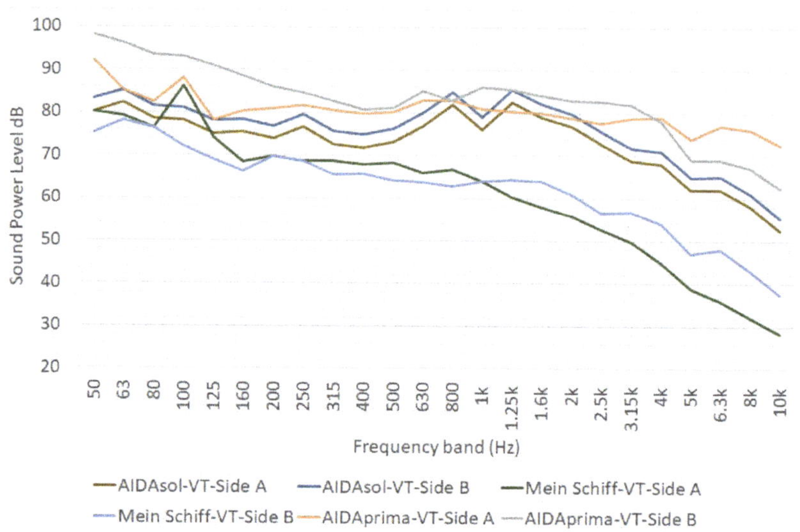

Figure 9. One-third octave sound power spectra of ventilation openings on the two sides of three vessels, as reported in [20]. L_W is expressed in dB for point sources, in dB/m for linear sources and in dB/m^2 for area sources.

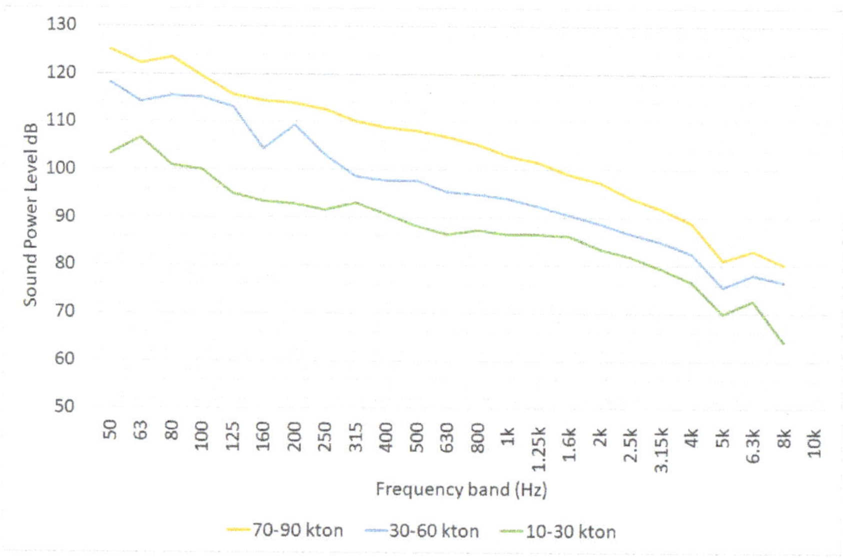

Figure 10. One-third octave sound power spectra of cruise ships divided into big (70 to 90 ktons), medium (30 to 60 ktons) and small size (10 to 30 ktons), as reported in [24]. L_W is expressed in dB for point sources, in dB/m for linear sources and in dB/m^2 for area sources.

A recent report on the noise emission of moored ships was issued by Tecnalia in 2018 [27]. The report was focused on evidencing the reduction of noise impacts thanks to cold ironing systems. In addition, the report contains useful indications about the noise emission of several kinds of moored ships based on noise measurements performed in Spanish ports in 2017; these noise data are reported in Table 7.

Table 7. Noise emission data of moored ship, as reported in [27]. * TEU: twenty-foot equivalent unit.

Type	Year	Size (GT)	Size (TEU *)	Reefer (TEU *)	Auxiliary Engine Power (kW)	Operating Conditions (kW)	Auxiliary Engine		Additional Source: Ventilation	
							L_w (dB(A))	Tonal Components/Low Frequency (dB)	Sound Power Level (dB(A))	Tonal Components/Low Frequency (dB)
Ro-Pax	2003	22,382	-	-	4200	900	109.3	6/0	113.2	6/0
Ro-Ro	1999	12,076	-	-	2 × 980	400	107.5	3/3	109.0	6/0
	2002	14,241	1129	153	-	-	97.4	3/6	-	-
	2008	7702	798	150	2 × 750	1 × 750	95.1	0/3	-	-
Containers	2007	8971	917	200	2 × 469	1 × 469	95.0	3/3	-	-
	2009	10,585	1036	-	-	-	90.2	0/3	92.3	3
	1973	28,372	-	-	2200	-	111.1	3/6	103.2	3/6
	2000	30,277	-	-	-	-	104.2	0/6	94.7	0/6
Cruise ship	2016	55,254	-	-	-	-	101.6	3/6	97.5	3/6
	2002	139,570	-	-	-	-	105.3	3/6	98.7	3/6
	2008	154,407	-	-	-	-	104.5	0/6	96.2	0/6

Even if it does not contain sound power data, it is worth considering the operational indications provided by the 2019 report of the Lloyd's Register on moored ship noise modelling [26]. The document suggests that the noise modelling procedure should consider the screening, reflection and absorption procedure of the ship's structure and should be performed at least in the 31.5–8000 Hz range. The inclusion of the following noise sources is recommended:

- Funnels and other exhaust stacks;
- Ventilation air intakes and exhaust;
- External fans;
- Hull radiated noise (if relevant);
- Cranes, pumps and any other equipment in operation.

They suggest considering the noise sources as single-point emitters, with the exception of large ventilation openings, which should be considered as surface noise source. The noise emission directivities of the noise sources have to be considered.

It may be of interest to mention the different approach used by Moro [40]; the noise emission of 290 meters and a 110,000 DWT cruise ship was modelled through software based on the beam method. The ship was defined through a 3D geometry model and all its noise sources were detected and characterized in terms of the sound power level. These emissions were defined using the procedures in compliance with the ISO 3744 standard. A comparison between the outcomes of this model with a noise measurement campaign was carried out and the outcomes showed an adequate agreement.

Finally, it is worth nothing that the FP7 SILENV project [30] defined two different methods to assess the noise emission of moored ships, as is reported in detail in the deliverable 5.2 of the project, "Noise and vibration label proposal" [31,32]. Each moored ship has to be modelled through a group of point, linear and area noise sources. Each one of these sources represents the relevant noise emitters of a moored ship such as funnels, intake and outlets of ventilation, and HVAC, etc.

3.7. Ro-Ro and Ro-Pax Ramp

The noise caused by the passage of vehicles on the Ro-Ro and Ro-Pax ramps can be relevant in a noise mapping project. Unfortunately, few works have been dedicated to its noise assessment. The EFFORTS project ([17,18]) allows us to give an estimation of these noise events in terms of sound power level spectra (Table 8). The assessment of noise emission was carried out through measurements on three ramps, one in Turku and two in Dublin. In the Turku assessment, the movement of goods was made through tractors; in the Dublin assessments, the ramps were used directly by trucks. Data reported in Table 8 shows that the noise emission data obtained from this study were subjected to a high degree of variability.

Table 8. Sound power level of Ro-Ro ramps, as reported in deliverable 2.4.3 of [18]. * The ship loading considers the whole measurement period, while other data consider the single event.

Ship Source	L_W (dB)	L_W (dB(A))
Ro-Ro ramp in Turku: tractor with trailer	114 ± 3	109 ± 3
Ro-Ro ramp in Turku: tractor without trailer	112 ± 4	106 ± 4
1st Ro-Ro ramp in Dublin: ramp noise	119 ± 5	112 ± 5
1st Ro-Ro ramp in Dublin: ship loading *	115 ± 2	108 ± 2
2nd Ro-Ro ramp in Dublin: ramp noise	121 ± 6	115 ± 6
2nd Ro-Ro ramp in Dublin: ship loading *	116 ± 2	109 ± 2
Ro-Ro ramp in Turku: tractor with trailer	114 ± 3	109 ± 3

The spectra of sound power levels from Table 8 are reported in Figure 11. It is worth noting that the spectra of the two ramps in Dublin are sensibly different; this is probably caused by the peculiarity of this noise source. Ramp noise emission is caused by the bumps between the ramp and the ground that happen when a track or a tractor passes over it. Each bump causes noise emissions that are considerably different from the others; there are a lot of factors influencing this (ground and ramp typology, tractors or truck velocity, weight of the tractors, etc.). In noise simulation activities, it is recommended to perform some devoted noise measurements on ramps similar to the ones to be modelled. Only if this is not possible, should this data reported in Figure 11 and Table 8 be used.

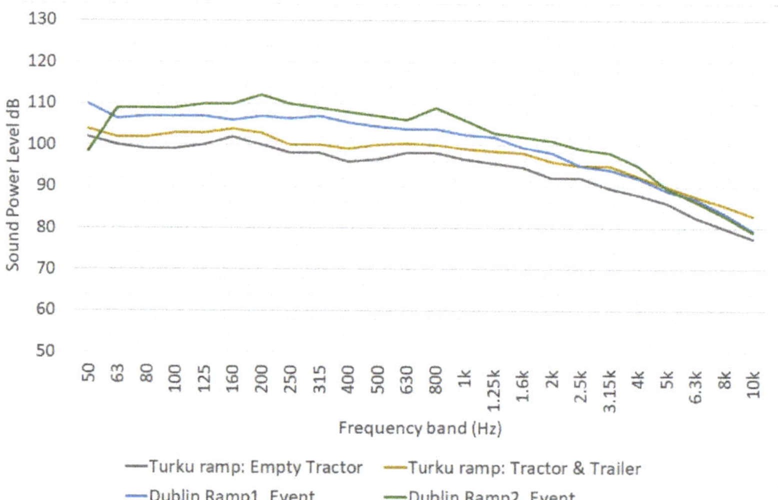

Figure 11. One-third octave sound power spectra of Ro-Ro ramps, as reported in deliverable 2.4.3 of [18]. L_W is expressed in dB for point sources.

3.8. Forklifts

The Green Cruise Port project [20] provided an overview of the noise emission of forklifts in a port area (Table 9). However, the sound power level of these devices can be retrieved from many databases, because they are also used outside port areas and their manufacturers are obliged to explicitly declare this data according to the 2000/14/EC directive. For instance, the SoundPLAN 8.2 database includes spectra for several kind of forklift, both under idling and in working conditions [41].

Table 9. Sound power level of forklifts, as reported in [20].

Source Description	L_W (dB(A))	Data Taken from
Small/medium diesel forklift	97	Manufacturer
Heavy duty size diesel forklift	107	Manufacturer
Electric forklift	90	[42]
Mobile crane for cruise ship	107	[43]

3.9. Seagoing Ships

In the harbor context, ships can represent an important source of noise both when berthed and when under navigation or during maneuvers. Navigation in the harbor is always at a very low speed; nevertheless, all the noise sources present during navigation are active such as air inlets and outlets, machinery noise, funnel noise, cargo treatment, etc. For now, measurement standards are available for vessels in inland waterways [44] and for small pleasure crafts [45], while no specific standard for seagoing ships is available. As a consequence, the scientific community has followed different approaches in measuring and reporting their results [46,47]. In [48], eight transits of two ferries (A, B), considering arriving and leaving with a source–receiver distance of about 170 m, were measured. Results are reported in Table 10 in terms of A-weighted SEL.

Table 10. A-weighted SEL of seagoing ships, as reported in [48].

Ship Type	Direction	SEL (dB(A))
A	Arriving	87–90
A	Leaving	89–90
B	Arriving	88–91
B	Leaving	88–89

No spectra were reported in the original work for a single ship, but a spectrogram of a single ship passage highlights the presence of string noise components at lower frequencies.

In [49], several measurements were carried out of a ferry during maneuvers in the port of Naples. The sound level meters were placed in two different locations: one on the short berth side and one on the long berth side. No spectral results were provided.

Fredianelli, et al. [29] performed the most complete effort to define the noise emission of different categories of seagoing ships by means of long-term pass-by measurements in the port of Livorno. The average sound power levels and spectra of different categories of big ships are reported in Table 11. Among the categories are: Ro-Ro, container, oil tanker, chemical tanker and ferry. The values for ferries are taken for a subsequent study by the same authors [50], focused only on ferries and the effects of parameters, such as ship speed and draught, and the distance from the microphone, have on the measured noise. Table 11 also includes the sound emissions of small and medium vessels performed by Bernardini, et al. in [11]. According to the paper, small vessels are meant to be small motorboats, sailing boats, and rigid-hulled inflatable boats, while medium vessels are small and mid-sized fishing boats, fireboats, and public security boats. Data reported in Table 11 for small and medium vessels are those reported in the original paper as "medium speed", 14.4–19.2 km/h (7.8–10.4 kn) for small vessels and 9.2–10.7 km/h (5.0–5.7 kn) for medium vessels.

Results consistently show the typical presence of strong noise components at lower frequencies for each ship type.

Table 11. One-third octave sound power spectra and sound power level of seagoing ships, as reported in [11,29,50]. Uncertainties can be retrieved from the original documents.

Ship Type	L_W dB(A)	20 Hz	25 Hz	31.5 Hz	40 Hz	50 Hz	63 Hz	80 Hz	100 Hz	125 Hz	160 Hz	200 Hz	250 Hz	315 Hz	400 Hz	500 Hz	630 Hz	800 Hz	1 kHz	1.25 kHz	1.6 kHz	2 kHz	2.5 kHz	3.15 kHz	4 kHz	5 kHz	6.3 kHz	8 kHz	10 kHz	12.5 kHz	16 kHz	20 kHz
															One-Third Octave Frequency Bands (dB(A))																	
Ro-Ro	87.5	99.2	98.3	96.7	97.7	96.9	97.1	94.8	91.4	90.0	88.8	86.2	84.6	83.1	81.2	80.1	78.7	77.2	75.9	74.1	72.5	70.9	68.8	67.1	65.6	64.8	64.5	64.4	67.0	68.6	69.2	69.2
Container	89.0	92.7	91.7	93.5	97.9	98.1	96.7	93.6	92.3	89.5	88.5	87.3	85.1	83.8	83.8	82.5	81.4	79.8	78.1	76.2	74.3	72.3	70.7	69.1	67.6	65.5	64.7	64.3	67.4	67.8	67.7	67.0
Oil tanker	82.6	86.5	85.9	88.3	90.5	89.0	87.3	89.5	87.6	84.2	80.3	78.5	78.0	76.7	75.1	74.2	73.1	72.4	71.5	70.9	69.3	67.8	68.6	66.6	64.7	64.0	64.4	63.6	65.0	66.9	67.6	67.5
Chemical tanker	85.9	87.1	89.5	90.6	94.4	95.0	91.8	91.1	89.2	88.3	84.8	83.2	84.5	81.3	79.8	78.6	77.3	75.9	75.4	73.1	71.8	69.6	67.8	67.0	65.0	63.7	62.6	62.1	64.1	65.3	66.1	67.5
Ferry	83.1	92.3	93.6	95.5	97.7	95.0	92.3	87.4	87.6	83.3	81.2	79.1	77.3	75.9	76.3	76.4	73.2	74.3	73.0	72.5	70.5	68.7	66.1	64.2	62.4	61.5	59.9	59.5	61.3	62.4	62.4	62.8
Small vessels	77.4	77.6	78.7	78.8	81.2	85.2	80.6	78.6	80.4	76.8	76.3	73.5	72.7	70.5	69.5	69.0	68.1	67.4	66.9	66.4	65.6	64.4	63.3	62.1	61.0	59.6	57.8	57.0	57.5	56.7	54.0	49.2
Medium vessels	83.5	78.1	80.4	85.5	86.3	90.9	85.6	87.3	89.7	87.4	84.4	82.1	80.9	77.4	75.5	75.4	75.1	73.8	72.6	70.9	69.6	68.0	66.9	65.8	64.2	62.3	60.4	59.8	60.3	59.6	56.6	51.7

3.10. Other Sources

Port areas can also be occupied by more common noise sources such as roads, rail, industries, power plants, waste treatment plants, etc. The assessment of the noise emission of these activities can be carried using the available CNOSSOS models [14]. Noise emission from parking can be evaluated using, for instance, the model developed by [51,52]. The assessment of the noise emitted from leisure activities is outside of the scope of this paper.

4. Conclusions

A comprehensive database of typical sound power levels and spectra of the various noise sources acting in ports was developed in the present paper. The sound power levels and spectra reported and summarized in the present work come from the research on both the scientific and technical literature, as well as from the deliverables and results of several European projects. These projects were developed on the framework of different programs (i.e., FP7, Interreg, ENPI CBC-MED, LIFE). Among them, the most important projects to be mentioned for having produced the higher amount of data are ANCHOR, REPORT and SILENV. As some of the authors participated in more than one of these projects, albeit with different aims of the funding programs and of the projects themselves, working in a sort of cluster of projects dealing with the same topic led to the development of knowledge on the topic that allowed the understanding of several different issues about port noise and the finding of solutions that are able to tackle them. One of the key issues was, in fact, the difficulty of developing environmental noise simulations of port noise leading to accurate and repeatable strategic noise maps and the subsequent noise action plans, as requested by the Environmental Noise Directive of the European community. In general, a scarcity of data was found for all sources considered. This may be due to a certain lack of standards and normative frameworks specific to these kinds of problems. In particular, the largest sources (such as transtainers and container ships) are those for which the lower amount of data is present. Moreover, it has to be underlined that sometimes the data sources avoid specifying whether the characterization measurements were made in compliance with a standard; this leads to a non-uniformity that makes difficult to compare results and to assess the reliability of the data presented. New guidelines aimed at tackling these issues were developed within the framework of the already cited REPORT project, recently ended, and it is foreseeable that these guidelines will be enacted in the future in order to have reliable, comparable and consistent measurements regarding port noise sources.

Such an organized dataset would be important for the present state of the scientific and technical literature as the complexity of noise produced by port infrastructure is high, with many different sources acting simultaneously. In recent years, port noise has been the object of some studies, but mostly those investigating only a singular type of noise source. Thus, a database comprising all of the spectral data retrievable was missing in the literature to the best of the authors' knowledge and it would be beneficial for technicians who produce noise maps or for other scientists willing to further improve a topic that still deserves lot of attention.

The development of such a database will act as a base for developing reliable numerical simulations in order to comply with the evermore restrictive standards and normatives regarding environmental pollution and the sustainability of transportation infrastructure.

Author Contributions: Conceptualization, S.S., D.B., F.D. and L.F.; methodology, S.S.; validation, S.S., D.B. and L.F.; formal analysis, S.S.; investigation, S.S., D.B., F.D., L.F. and T.G.; resources, C.S. and G.B.; writing—original draft preparation, S.S., D.B., F.D., L.F. and T.G.; writing—review and editing, S.S., D.B., F.D., L.F. and T.G.; supervision, C.S. and G.B.; project administration, C.S. and G.B.; funding acquisition, C.S. and G.B. All authors have read and agreed to the published version of the manuscript.

Funding: This research was funded by the EU ANCHOR LIFE project—LIFE17 GIE/IT/000562.

Data Availability Statement: Not applicable.

Conflicts of Interest: The authors declare no conflict of interest.

Appendix A

Table A1. Standards used in the acoustic measurements carried out to define the sound power level of port noise sources.

Reference	Port Noise Source	Measurement Carried out in Compliance with
REPORT Project [16]	Transtainer (Section 3.1)	NA
	Reach stacker (Section 3.2)	NA
	Gantry cranes (Section 3.4)	NA
	Reefer (Section 3.5)	NA
EFFORTS Project [17,18]	Transtainer (Section 3.1)	ISO 3744 *
	Reach stacker (Section 3.2)	ISO 3744 *
	Straddle carrier (Section 3.3)	ISO 3744 *
	Gantry cranes (Section 3.4)	ISO 3744 *
	Reefer (Section 3.5)	ISO 3744 *
	Moored ship (Section 3.6)	ISO 3744 *
	Ro-Ro and Ro-Pax ramps (Section 3.7)	ISO 3744 *
Witte, J. [19,22]	Straddle carrier (Section 3.3)	NA
	Gantry cranes (Section 3.4)	NA
	Moored ships	NA
Danish Ministry of the Environment [25]	Reefer (Section 3.5)	NA
	Moored ships (Section 3.6)	NA
GREEN CRUISE PORT [20]	Reefer (Section 3.5)	ISO 3746
	Engine of moored ships (Table 4 of Section 3.6)	ISO 9614-2
	Ventilation of moored ships (Table 5 of Section 3.6)	NA **
	Funnels of moored ships (Figure 8 of Section 3.6)	DIN 45 635-47:1985
	Ventilation openings of moored ships (Figure 8 of Section 3.6)	ISO 3746
	Diesel forklift (Section 3.7)	NA **
	Electric forklift (Section 3.7)	NA
NEPTUNES Project [21]	Reefer (Section 3.5)	NA
	Moored ships (Section 3.6)	NA
Tecnalia, [27]	Engine of Moored ships (Section 3.6)	ISO 3746 *
	Ventilation of Moored ships (Section 3.6)	ISO 3746 *
Di Bella and Remigi [24]	Moored ships (Section 3.6)	ISO 8297, ISO 3744 and ISO 3746
Moro [40]	Moored ships (Section 3.6)	ISO 3744
Di Bella et al. [48]	Seagoing ships (Section 3.9)	UNI 11143
Fredianelli et al. [29]	Seagoing ships (Section 3.9)	BS EN ISO 2922:2000 + A1:2013

* The authors declare that measurements were carried out via a simplified application of the standard. ** Data directly supplied by manufacturer.

References

1. Review of Maritime Transport. UNCTAD/RMT/2021. Available online: https://unctad.org/system/files/official-document/rmt2021_en_0.pdf (accessed on 3 August 2022).
2. Container throughput at Ports Worldwide from 2012 to 2021 with a Forecast for 2022 through 2025. Available online: https://www.statista.com/statistics/913398/container-throughput-worldwide/ (accessed on 3 August 2022).
3. 2021 Worldwide Cruise Line Market Share. Available online: https://cruisemarketwatch.com/market-share/ (accessed on 3 August 2022).
4. Merk, O. The Competitiveness of Global Port-Cities: Synthesis Report. OECD Regional Development Working Papers 2013/13. Available online: https://www.oecd.org/regional/the-competitiveness-of-global-port-cities-9789264205277-en.htm#:~{}:text=Ports%20and%20cities%20are%20historically,localised%20in%20the%20port%2Dcity (accessed on 3 August 2022).

5. Trozzi, C.; Vaccaro, R. Environmental Impact of Port Activities. In *Maritime Engineering and Ports II*; Brebbia, C.A., Olivella, J., Eds.; WIT Press: Cambridge, MA, USA, 2000.

6. Badino, A.; Borelli, D.; Gaggero, T.; Rizzuto, E.; Schenone, C. Analysis of Airborne Noise Emitted from Ships. In *Sustainable Maritime Transportation and Exploitation of Sea Resources*; Rizzuto, E., Guedes Soares, C., Eds.; Taylor & Francis Group: Abingdon, UK, 2012.

7. Fredianelli, L.; Bolognese, M.; Fidecaro, F.; Licitra, G. Classification of noise sources for port area noise mapping. *Environments* **2021**, *8*, 12. [CrossRef]

8. Murphy, E.; King, E.A. An assessment of residential exposure to environmental noise at a shipping port. *Environ. Int.* **2014**, *63*, 207–215. [CrossRef] [PubMed]

9. Badino, A.; Borelli, D.; Gaggero, T.; Rizzuto, E.; Schenone, C. Control of Airborne Noise Emissions from Ships. In Proceedings of the International Conference on Advances and Challenges in Marine Noise and Vibration 21 MARNAV 2012, Glasgow, Scotland, UK, 5–7 September 2012.

10. World Health Organization. Environmental Noise Guidelines for the European Region. 2019. Available online: https://www.who.int/europe/publications/i/item/9789289053563 (accessed on 3 August 2022).

11. Bernardini, M.; Fredianelli, L.; Fidecaro, F.; Gagliardi, P.; Nastasi, M.; Licitra, G. Noise assessment of small vessels for action planning in canal cities. *Environments* **2019**, *6*, 31. [CrossRef]

12. NoMEPorts Project, Good Practice Guide on Port Area Noise Mapping and Management. 2008. Available online: https://www.ecoports.com/assets/files/common/publications/good_practice_guide.pdf (accessed on 3 August 2022).

13. Fredianelli, L.; Gaggero, T.; Bolognese, M.; Borelli, D.; Fidecaro, F.; Schenone, C.; Licitra, G. Source characterization guidelines for noise mapping of port areas. *Heliyon* **2022**, *8*, e09021. [CrossRef]

14. Kephalopoulos, S.; Paviotti, M.; Anfosso-Lédée, F. *Common Noise Assessment Methods in Europe (CNOSSOS-EU)*; Publications Office of the European Union: Luxembourg, 2012. [CrossRef]

15. IEC. *IEC 61672-1:2013 Electroacoustics—Sound Level Meters—Part 1: Specifications*; International Electrotechnical Commission: Geneva, Switzerland, 2013.

16. REPORT Project. Available online: https://interreg-maritime.eu/web/report/checosarealizza (accessed on 3 August 2022).

17. Hyrynen, J.; Maijala, V.; Melin, V. Noise Evaluation of Sound Sources Related to Port Activities. In Proceedings of the Euronoise 2009, Edinburgh, UK, 26–28 October 2009.

18. EFFORT Project. Available online: http://efforts-project.tec-hh.net/index.html (accessed on 3 August 2022).

19. Witte, J. Container Terminals and Noise. In Proceedings of the Internoise 2008, Shanghai, China, 26–29 October 2008.

20. GREEN CRUISE PORT Project. Available online: http://www.greencruiseport.eu/Home.html (accessed on 3 August 2022).

21. NEPTUNES Project. Available online: https://neptunes.pro/ (accessed on 3 August 2022).

22. Witte, J. Noise Emission Ro-Ro Terminals. In Proceedings of the Euronoise 2009, Edinburgh, UK, 26–28 October 2009.

23. Badino, A.; Borelli, D.; Gaggero, T.; Rizzuto, E.; Schenone, C. Airborne noise emissions from ships: Experimental characterization of the source and propagation over land. *Appl. Acoust.* **2016**, *104*, 158–171. [CrossRef]

24. Di Bella, A.; Remigi, F. Evaluation and control of cruise ships noise in urban areas. In Proceedings of the ICSV20, Bangkok, Thailand, 7–11 July 2013.

25. Noise from Ships in Ports: Possibilities for Noise Reduction, Environmental Project No. 1330. 2010. Available online: https://mst.dk/media/mst/66165/978-87-92668-35-6.pdf (accessed on 3 August 2022).

26. Lloyd's Register. Procedure for the Determination of Airborne Noise Emissions from Marine Vessels. 2019. Available online: https://www.lr.org/en/shipright-procedures/#accordion-additionaldesign&constructionprocedure(adp) (accessed on 3 August 2022).

27. Tecnalia. Assessment of the Acoustic Benefit of the Power Supply to Ships Moored in Ports (Cold Ironing). Available online. 2018. (accessed on 3 August 2022).

28. Santander, A.; Aspuru, I.; Fernandez, P. OPS Master Plan for Spanish Ports Project. Study of Potential Acoustic Benefits of On-shore Power Supply at Berth. In Proceedings of the Euronoise 2018, Crete, Greece, 27–31 May 2018.

29. Fredianelli, L.; Nastasi, M.; Bernardini, M.; Fidecaro, F.; Licitra, G. Pass-by characterization of noise emitted by different categories of seagoing ships in ports. *Sustainability* **2020**, *12*, 1740. [CrossRef]

30. FP7 SILENV Project. Available online: https://cordis.europa.eu/project/id/234182/it (accessed on 3 August 2022).

31. Deliverable 5.2 of the SILENV Project. "Noise and Vibration label proposal". 2012.

32. Borelli, D.; Gaggero, T.; Rizzuto, E.; Schenone, C. Holistic control of ship noise emissions. *Noise Mapp.* **2016**, *3*. [CrossRef]

33. Borelli, D.; Gaggero, T.; Rizzuto, E.; Schenone, C. Measurements of Airborne Noise Emitted by a Ship at Quay. In Proceedings of the ICSV22, Florence, Italy, 12–16 July 2015.

34. Curcuruto, S.; Marsico, G.; Atzori, D.; Mazzocchi, E.; Betti, R. Environmental Impact of Noise Sources in Port Areas: A Case Study. In Proceedings of the ICSV22, Florence, Italy, 12–16 July 2015.

35. Marsico, G. Infrastrutture di Trasporto Portuali: Metodologie e Procedure per la Valutazione e il Contenimento dell'Impatto Acustico Sull'Ambiente. Ph.D. Thesis, Università di Roma Sapienza, Dipartimento di Ingegneria Astronautica, Elettrica ed Energetica, Rome, Italy, 2007. (In Italian)

36. Emission Data of Port Noise Sources in Source 01dB Software Database. Original data from DGMR-report W.92.03.59.C d.d.

37. Stampe, O.B. *Lyd i VVS-anlæg*; Skarland Press AS: Rakkestad, Norway, 1998.

38. *ISO 3746:2010*; Acoustics—Determination of Sound Power Levels and Sound Energy Levels of Noise Sources Using Sound Pressure—Survey Method Using an Enveloping Measurement Surface over a Reflecting Plane. ISO: Geneva, Switzerland, 2010.

39.   Draganchev, H.; Valchev, S.; Pirovsky, C. Experimental and Theoretical Research of Noise Emitted by Merchant Ships in Port. In Proceedings of the ICSV19, Vilnius, Lithuania, 8–12 July 2012.
40.   Moro, L. Setting of on-Board Noise Sources in Numerical Simulation of Airborne Outdoor Ship Noise. In Proceedings of the 9th Youth Symposium on Experimental Solid Mechanics, Trieste, Italy, 7–10 July 2010.
41.   SoundPLAN GmbH. Emission Data of Port Noise Sources in SoundPLAN Software Sersion 8.2. Available online: http://www.soundplan.eu/english (accessed on 3 August 2022).
42.   Umweltbundesamt. Forum Schall: Emissionsdatenkatalog 2016. Available online: https://www.oal.at/images/Forum_Schall/Arbeitsbehelfe/Emissionsdatenkatalog_2022.pdf (accessed on 3 August 2022).
43.   LAIRM CONSULT GmbH. *Schalltechnische Untersuchung für das Geplante Cruise Center 3 in Hamburg-Steinwerder*; Hamburg Port Authority: Hamburg, Germany, 2013.
44.   *ISO 2922:2020*; Acoustics—Measurement of Airborne Sound Emitted by Vessels on Inland Waterways and Harbours. ISO: Geneva, Switzerland, 2020.
45.   *ISO 14509-1:2008*; Small Craft—Airborne Sound Emitted by Powered Recreational Craft—Part 1: Pass-by Measurement Procedures. ISO: Geneva, Switzerland, 2008.
46.   Di Bella, A. Evaluation Methods of External Airborne Noise Emissions of Moored Cruise Ships: An Overview. In Proceedings of the 21st International Congress on Sound and Vibration 2014, ICSV 2014, Beijing, China, 13–17 July 2014; Volume 2, pp. 964–971.
47.   Fausti, P.; Santoni, A.; Martello, N.; Guerra, M.C.; Di Bella, A. Evaluation of Airborne Noise due to Navigation and Manoeuvring of Large Vessels. In Proceedings of the 24th International Congress on Sound and Vibration, ICSV 2017, London, UK, 23–27 July 2017.
48.   Di Bella, A.; Tombolato, A.; Cordeddu, S.; Zanotto, E.; Barbieri, M. In Situ Characterization and Noise Mapping of Ships Moored in the Port of Venice. In Proceedings of the European Conference on Noise Control 2008, Paris, France, 29 June–4 July 2008; pp. 1949–1953.
49.   Coppola, T.; Mocerino, L.; Rizzuto, E.; Viscardi, M.; Siano, D. Airborne Noise Prediction of a Ro/Ro-Pax Ferry in the Port of Naples. Technology and Science for the Ships of the Future. In Proceedings of the NAV 2018, Trieste, Italy, 20–22 June 2018.
50.   Nastasi, M.; Fredianelli, L.; Bernardini, M.; Teti, L.; Fidecaro, F.; Licitra, G. Parameters affecting noise emitted by ships moving in port areas. *Sustainability* **2020**, *12*, 8742. [CrossRef]
51.   Bavarian State Agency for the Environment. *Parking Area Noise*, 6th ed.; Bavarian State Ministry for the Environment: Munich, Germany, 2007.
52.   Bayer. *Parkplatzlärmstudie, (Parking Lot Study)*, 6th ed.; Landesamt für Umwelt: Augsburg, Germany, 2007. (In German)

International Journal of
Environmental Research and Public Health

Article

Seafarers' Perception and Attitudes towards Noise Emission on Board Ships

Luka Vukić [1,*], Vice Mihanović [2], Luca Fredianelli [3] and Veljko Plazibat [1]

[1] Department for Maritime Management Technologies, Faculty of Maritime Studies, University of Split, Ruđera Boškovića 37, 21000 Split, Croatia; veljko.plazibat@pfst.hr
[2] Port Authority Split, Gat Svetog Duje 1, 21000 Split, Croatia; vice.mihanovic@gmail.com
[3] Physics Department, University of Pisa, Largo Bruno Pontecorvo 3, 56127 Pisa, Italy; fredianelli@df.unipi.it
* Correspondence: lvukic@pfst.hr; Tel.: +385-985-498-49

Abstract: Noise has long been neglected as an environmental pollutant and impairment health factor in maritime transport. Recently, acoustic pollution indicates the highest growth in transport external cost unit values. In 2020, questionnaires were submitted to seafarers to examine their noise exposure and perception on board and attitudes towards noise abatement measures. Responses of 189 participants were processed using descriptive statistics and Likert scale valuation, while their consistency was tested with indirect indicators using linear regression and correlation test. Results show that more than 40% of respondents do not consider noise as a significant environmental problem. The negative perception among respondents with $\geq$10 years of work experience was much lower (23.53%). Most are aware of the onboard noise harmful effects that can influence their health. Despite that, they use personal protection equipment only sometimes. A higher positive perception was recorded in groups of respondents with a university degree (90%), work experience longer than ten years (82.35%), and monthly income higher than 4000 € (70%). Respondents are not strongly motivated to participate in funding noise mitigation measures, and such a viewpoint is not related to their monthly incomes. The low awareness and motivation regarding acoustic pollution generally shown by the surveyed seafarers should be watched as a threat by the company managers. Better education and awareness are likely to be crucial to change the current state of affairs.

Keywords: seafarers; acoustic pollution; noise onboard ship; health impact; environmental pollution; noise survey

Citation: Vukić, L.; Mihanović, V.; Fredianelli, L.; Plazibat, V. Seafarers' Perception and Attitudes towards Noise Emission on Board Ships. *Int. J. Environ. Res. Public Health* **2021**, *18*, 6671. https://doi.org/10.3390/ijerph18126671

Academic Editor: Julio Diaz

Received: 4 May 2021
Accepted: 18 June 2021
Published: 21 June 2021

Publisher's Note: MDPI stays neutral with regard to jurisdictional claims in published maps and institutional affiliations.

1. Introduction

The negative impact of transport on the environment and human health is usually expressed through external costs, where the noise cost has recently become a significant source of damage. These costs are not covered by the stakeholders of the logistics transport chain but are a burden to society. External cost is expressed as a price per unit of harmful transport product (e.g., decibels (dB) for noise). Based on the recent data on the external costs in transport retrieved from relevant literature [1–3], the noise external costs unit prices have increased more than 3.5 times in the last 12 years, an increase not recorded in any other external cost component in the sector. Reasons are changes in perception of noise pollution, modified regulations, insufficient and expensive protection measures, and stricter valorization due to recent findings of the noise impact on health. Recently, noise costs have become a significant factor in the transport impact on human health and the environment, accounting for almost 7% of total external transport costs in the European Union (EU) [3].

The World Health Organization (WHO) has recognized noise pollution as not only an environmental nuisance but a threat that can damage health and reduce the nearby property value [4]. More than 20% of EU residents have been exposing to an excessive noise level [5]. Prolonged exposure to noise levels above 55 dB(A) can be detrimental to health, while levels above 65 dB(A) should not be tolerated [6] over the long term. The health effect of noise starts from the "indirect" ones, such as annoyance (nuisance), sleep disturbance, stress, anxiety occurring at lower levels of exposure, and "direct effects" when the exposure exceeds 85 dB(A). Direct effects include tinnitus, cognitive impairment in children, ischemic heart disease, and hypertension [7]. Also, for these reasons, noise has been recognized as one of the main reasons for the reduced life quality in urban and country areas [8,9].

The transportation sector is the principal cause of environmental noise, where road contributes to 65%, air to 20%, and railway to 15% of the overall level of noise impact in the environment [8]. Maritime and inland waterways transports have a reduced significance [3] with the consequence that few studies have been published in the scientific literature. However, ship noise onboard can endanger seafarers and passengers, while underwater and airborne emitted ship noise can affect port areas and coastal residents, even the fauna on maritime routes [10]. Based on the research of [11–14], the principal source of noise on board can be assumed to be the engine room, where the highest levels of intensity can be found. On most ships, noise levels over 100 dB(A) are present, reach the levels of 110 dB(A) in the noisier area and decrease depending on the location on board. Permanent and simultaneous exposure to noise, vibration, and heat on ships contributes significantly to developing anxiety in seafarers [11]. Noise exposure onboard increases mobility during sleep by 12%, and conjoined with other agents like caffeine and nicotine, may cause shallow sleep [15]. A better rest improves health and safety, which indirectly reduces the frequency of onboard accidents and improves productivity [16]. There is still debate about the relationship between ship noise and arterial hypertension occurrence in seafarers [17]. Hearing loss is a leading occupational disease, and seafarers working in an engine room on a ship are particularly at risk [18]. The Norwegian Centre for Maritime Medicine reviewed noise levels on board and their influence on seafarers [19]. Nastasi et al. [20] point out that noise has only recently been taken into account in the port sustainability assessment. Exposure of citizens to the noise in port areas has also been underestimated [21]. In the port of Livorno, e.g., during arriving and departing ships, the noise increases by 6–10 dB above the existing background noise [22]. Witte [23] states that the mitigating measures of ship noise at berth, like shore power connection, can drastically improve air quality but not reduce noise emission proportionally.

In 2012, the International Maritime Organization (IMO) adopted the Convention for the Safety of Life at Sea (SOLAS) with a requirement for noise reduction, both by adequate solutions in ship construction and personal protection equipment for seafarers following The Code on noise levels on board ships [24]. The Code has been developed to provide international standards for protection against noise and tools to promote "hearing saving" environment onboard ships. Unfortunately, not enough public awareness of the harmfulness of noise on ships and in ports [25] has been raised since then. Raising awareness and education about the harmful effects of noise is crucial, and such initiatives come from all over [26]. Despite regulations, the intensity of noise on ships often exceeds the permissible values determined by Directive 2003/10/EC [12,13,27]. There is also relatively little interest in the scientific community, and papers on noise as working environment and barrier to development are not frequent.

When exposed to environmental noise levels between 50 and 75 dB(A), noise experience and acceptance vary on individual. Also, the noise tolerance threshold is determined independently, as one can tolerate higher noise intensities while another cannot tolerate noises below 50 dB regardless of education on the detrimental effects of noise. This aspect led scientists to introduce the term noise sensitivity. It is a measuring unit of non-auditory influence of the environmental noise, which is individually different at the same intensity noise exposure [28]. Some other adverse factors have collateral effects on noise perception, such as meteorological conditions or, in general, changing conditions at the site of perception. Therefore, valorization based on statement, impression, attitude, and opinion is imprecise and uncertain, and the possibility of objectifying disorders is limited.

The present paper aims to determine the seafarers' noise pollution perception on board and evaluate their attitudes towards noise exposure. The aim is reached using a structured questionnaire based on collecting general noise perception data on environment and health, as well as noise perception on board and in place of residence. Encouraged by the current trend and sudden increase in the external noise costs, the research would contribute to the topic's actuality. Noise cost marginalization in maritime transport refers only to the low capital share and does not to the real significance of noise pollution. The research also wishes to contribute and drive the education and raising seafarers' awareness of the noise harmfulness on board. Awareness level about the harmfulness of noise in people who are professionally exposed to it and therefore may suffer health consequences is a good indicator of how much significance is attached to noise as an environmental pollutant.

2. Materials and Methods

A structured questionnaire (Appendix A) for seafarers was composed about the perception and intensity of noise pollution in general and onboard ships. Noise analysis is combined with the top-down approach through the willingness to pay value (WTP), and alternatively, willingness to accept (WTA), multiplied by the number of noise-exposed persons to obtain average or total external noise costs [7,29]. Thus, the noise valorization is identified with the people's motivation in how much they are willing to spend for implementing the measures that will reduce the noise and, alternatively, how much compensation they claim for noise tolerance. Awareness of the harmful effects of noise is of great importance for conducting such a survey. When awareness of the noise exposure detrimental effects is not sufficient, a credible response can be obtained indirectly using a hedonic pricing method (HP). The method enables estimating one's attitudes towards noise pollution over his/her opinion on whether and to what extent noise affects own real estate prices and rental prices [7]. The present paper examined the seafarers' willingness to participate in financing noise abatement (WTP) as a good indicator of what extent an individual attaches importance to the topic. The respondents' objectivity was tested by questions about the need for a salary supplement due to noise exposure (WTA), perception of noise in own household, and noise impact on the own apartment value (HP). For the simple estimation of the noise intensity to which they are exposed, the respondents could use a decibel level comparison table attached in the questionnaire and choose the option. To some questions, respondents had to answer using the Likert scale. Data were processed using descriptive statistics. The correlation test (CORREL) and linear regression (LR) were used to determine the dependency between the size of monthly income (MI) and WTP as well as the requirement for a salary supplement due to noise impact (WTA) and WTP. The possible WTA and WTP values correlation with the estimations on the own apartment values loss due to noise (HP) were also determined. All calculations were made in spreadsheets. The methodological concept applied as sketched in Figure 1 aimed to objectify the consistency of the responses.

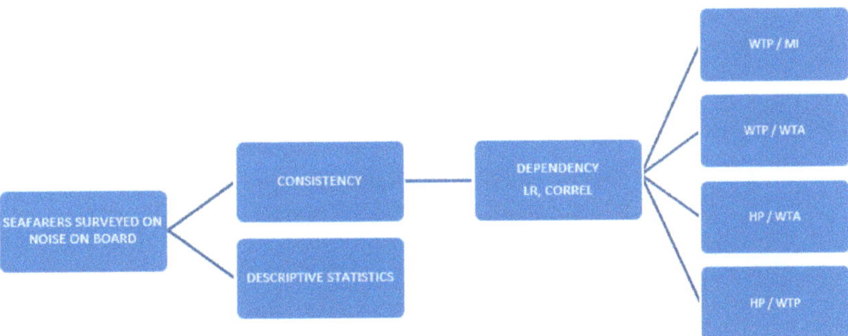

Figure 1. Flow chart of the research.

Expectations from respondents, who were occupationally exposed to noise pollution and aware of the harmful effects of noise on the environment and human health, are as follows: that those with higher monthly income will contribute more for noise mitigation (WTP/MI); that those who seek higher compensation for occupational noise pollution will contribute more for noise mitigation (WTP/WTA); that those who contribute more for noise mitigation also estimate the greater loss in value of their property due to noise (HP/WTP); that those who seek higher compensation for occupational noise exposure simultaneously estimate the corresponding loss in value of their property due to noise (HP/WTA).

In order to exclude subjectivity in the choice of answers, the F-test was used to examine the response dispersion differences to the noise perception at work and in their household. The same was examined in the groups of participants who indicated a possible leaving from the ship, respectively, changing the housing location due to noise exposure. All calculations were made in MS Excel.

The research was conducted from February to June 2020 at the Faculty of Maritime Studies in Split, Croatia. All respondents were participants in the course of additional education of seafarers (which is not related to a topic of noise). All respondents were Croatian citizens.

3. Results

In 2020, the questionnaire was applied to 189 seafarers with an average age of 35 years (27–52 years) and an average work experience of 11.5 years (4–29 years) with a median of 10 years (y). An average income was 3250 € a month (1000–5000 €). They work on merchant and passenger ships, being on board continuously for at least two months, followed by a month's rest on land. There were 171 male and 18 female seafarers in the research. The perception of respondents is shown in Table 1.

Table 1. Perception of the harmful effects of noise on the environment and health.

Perception	Environment		Health		
	pos	**neg**	**pos**	**pos/neg**	**neg**
General	58.73	41.27	53.97	26.98	19.05
Experience < 10 y	52.17	47.83	43.48	32.61	23.91
Experience ≥ 10 y	76.47	23.53	82.35	11.76	5.88
Secondary school	60.38	39.62	47.17	30.19	22.64
Bachelor/Master	50.00	50.00	90.00	10.00	0.00
Income (1.2)	61.11	38.89	50.00	25.00	25.00
Income (3)	61.90	38.10	40.00	40.00	20.00
Income (4.5)	57.14	42.86	70.00	15.00	15.00

The research results show that 41.27% of respondents do not consider noise pollution a significant environmental problem. Concerning education, almost the same percentage of the above perception was recorded among the respondents with secondary education (39.62%). It unexpectedly increased to 50% among those with higher education levels. Dispersion of respondents by the work experience in years is reported in Figure 2.

The median of 10 years was the criteria for creating comparative groups, a group <10 y, $n = 87$, and a group ≥10 y, $n = 92$. The variance examined with the two tail F tests shows statistically significant difference ($p = 3.09 \times 10^{-33}$, $\alpha = 0.05$). The negative perception among respondents with ≥10 years of work experience was much lower (23.53%) compared to respondents with <10 years of work experience (47.83%). Monthly income does not affect the perception of noise pollution. The statement that air pollution in maritime transport is a bigger problem than noise support 93.44% of the seafarers surveyed.

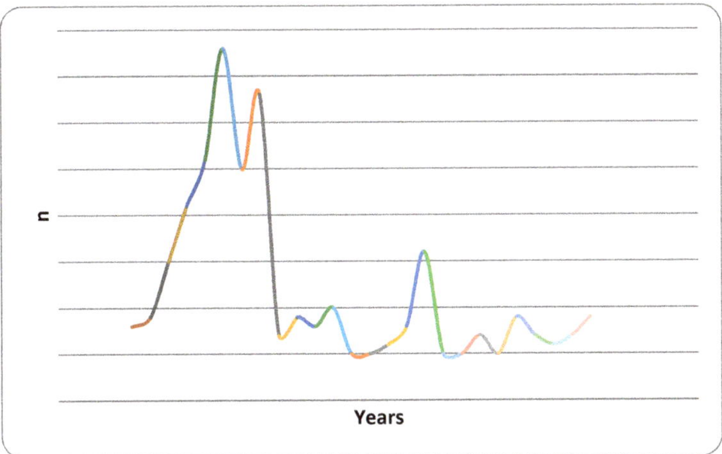

Figure 2. Dispersion of respondents by the work experience.

More than 50% of respondents are aware of the harmful effects of noise on health, more than 25% are aware of this at least partly, and 19% of respondents deny them. A higher positive perception was recorded in groups of respondents with a university degree (90%), work experience longer than ten years (82.35%), and monthly income higher than 4000 € (70%).

On a Likert scale, ranging from 1 to 5, respondents rated noise exposure on board as 3.85 (1 = does not interfere at all, 2 = interferes very little, 3 = little, 4 = much, 5 = very much), and equally during working hours (3.11) and rest periods (3.15). According to the attached intensity table, the estimated noise intensity during working hours is supposed at a range of 80–85 dB, and during rest hours at a range of 50–55 dB. The share of seafarers willing to provide salary supplement due to noise exposure was 5.75%. About 13.33% of respondents considered leaving the ship due to noise. On a Likert scale range from 1 to 3, the noise protection equipment use was at 2.37 (1 = never, 2 = sometimes, 3 = always). Vibration exposure on the same scale was rated with 2.22.

The surveyed seafarers indicated a willingness to pay an average of 65 € per year for noise mitigation. The dependence of the size of payments declared for noise mitigation on monthly incomes was examined by linear regression, as reported in Figure 3.

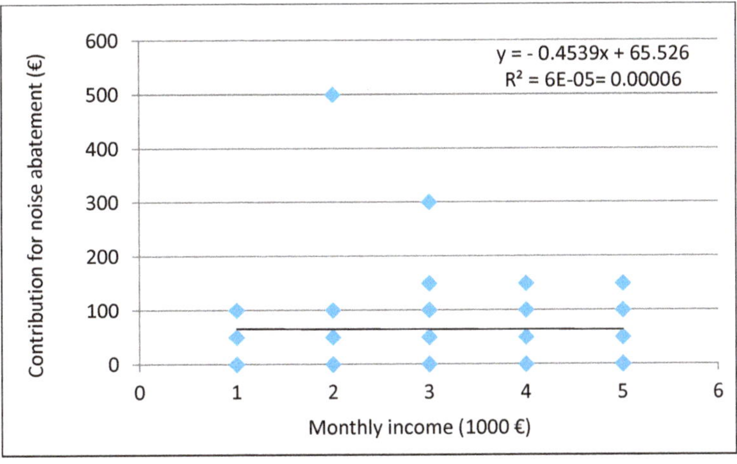

Figure 3. Dependence of declared contributions for noise abatement on monthly incomes.

The dependence between the given parameters was not determined ($R^2 = 0.00006$). The correlation test obtained value, r = 0.0075, confirms the absence of any relationship.

Furthermore, the dependence of the size of the payment declared for noise abatement on the request size for salary supplement due to noise was examined by linear regression as reported in Figure 4. Even this resulted not to be determined ($R^2 = 0.018$). The correlation coefficient r = 0.13398 indicates a very weak positive correlation.

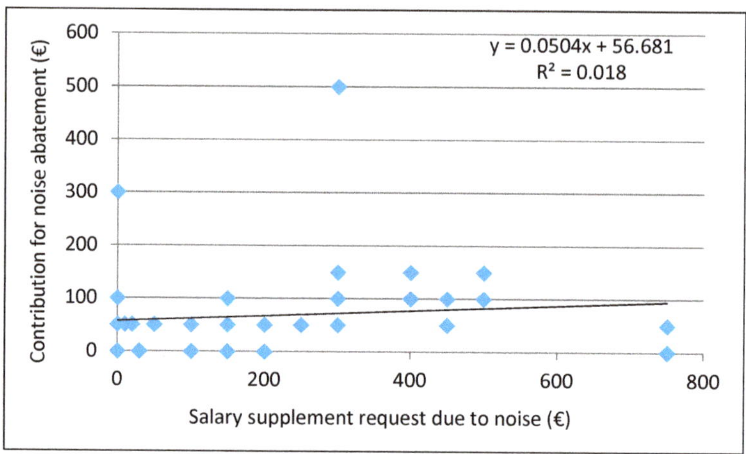

Figure 4. Dependence of annual contribution amount for noise abatement on requests for salary supplement due to noise.

On a Likert scale range from 1 to 5, respondents rated the perception of noise in their households with 2.27 (1 = does not interfere at all, 2 = interferes very little, 3 = little, 4 = much, 5 = very much), mostly at night (2.05 at a Likert scale range from 1 to 4 (1 = does not interfere at all, 2 = interferes at night, 3 = during the day, 4 = day and night). According to the attached table, they estimated the intensity of the household noise in the range between 50–55 dB by day and 35–40 dB at night. The surveyed seafarers believe that noise affects the value of the apartment by an average of 9.77%. Only 11.29% of respondents considered moving from their residence due to noise.

The variance differences in response groups on noise perception at the respondents' workplace and their homes were examined using the F test. The same procedure was applied to groups who declared intention to leave the workplace and move from their apartments due to noise, respectively. There were no statistically significant differences in variance among groups ($p = 0.2910$, one tail; $p = 0.1699$, one tail). The correlation test result, r = 0.1961, shows a very weak positive correlation between the last two groups.

The dependence of attitudes about the noise impact on own apartment value on those about the salary supplement request due to noise at the workplace was examined by linear regression, as reported in Figure 5. The low coefficient of determination ($R^2 = 0.0546$) indicates a minimal degree of dependence between the two groups of responses. The correlation value determined by the correlation test, r = 0.23363, shows a very weak positive correlation between the examined groups.

The same tests were used to find the dependence of attitudes towards the noise impact on the own apartment value on attitudes towards a voluntary contribution for noise abatement, as reported in Figure 6. The low coefficient of determination $R^2 = 0.0095$ and a correlation coefficient r = 0.0973 are found, indicating the absence of dependence and correlation between the settings.

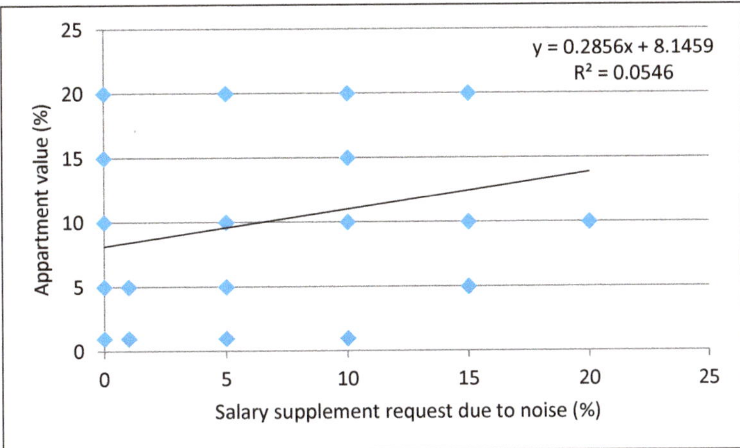

Figure 5. Dependence of attitudes about the noise influence on own apartment value on the amount of request for salary supplement due to noise.

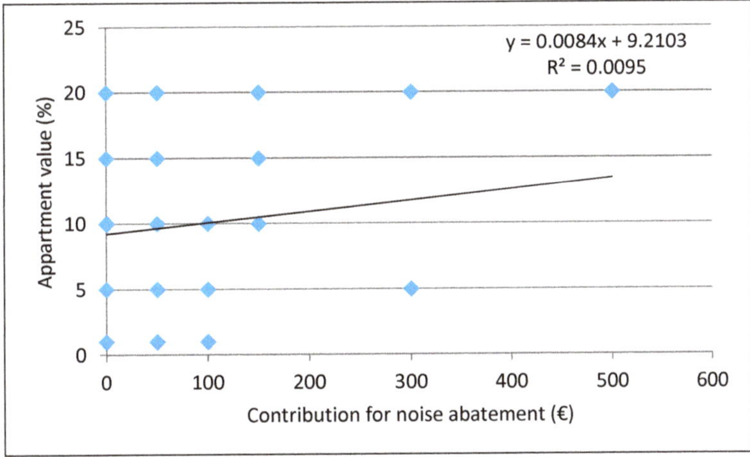

Figure 6. Dependence of attitudes about the noise influence on own apartment value on the declared contribution for noise abatement amount.

4. Discussion

The submitted questionnaires showed that almost half of the surveyed seafarers, in general, do not perceive onboard noise as a significant environmental problem in maritime transport, even if they are aware that prolonged noise exposure can have consequences for their health. According to the European Environmental Agency [30], this phenomenon happens to other people too. Subjective responses to noise depend not only on exposure levels but also on personality traits, expectations, and situational factors [31,32]. The results showed a noise harmfulness better perception in seafarers with more work experience, and noise health impact perception was also better in those with higher education and income. Choosing appropriate values, surveyed seafarers estimated their noise exposure level on board by the intensity that can damage their health and compromise their rest hours. The estimated average noise intensity during working hours was at almost 85 dB. This value follows findings obtained by Oldenburg et al. [11] and measured by Mansi et al. [14]. They are, obviously, insufficiently protected as they use noise protection agents only occasionally.

Despite the actual situation, seafarers are not ready to invest significant funds in noise mitigation, not even when it comes to their health. The amount of the declared financial contribution does not depend on the monthly income or whether they receive a monthly allowance for working in noise. This attitude objectifies the level of perception of noise pollution. The perception of noise in own apartment is consistent with the perception in the workplace. In general, respondents do not want to leave the workplace due to noise nor consider moving out of the apartment. Their attitudes to the need for noise reduction are inconsistent. The absence of any dependence of the amount of contribution for noise reduction on control indicators and control indicators on each other indicates other motives for such selection concerning the adopted attitudes about noise hazards. A similar conclusion has been published by Picu et al. [33]. Noise pollution has not sufficiently become aware among seafarers even though they are directly exposed to it in the workplace, contrary to air pollution, which they are more exposed to globally than locally. Insufficient education is probably the main reason for the weak perception of noise pollution among seafarers. A low level of perception by seafarers with a university degree could present a confirmation of this thesis. The lack of knowledge was the main reason for the port authorities' response to a special call for noise within the Interreg Maritime program [34].

The paper of Bernotaitė and Malinauskienė [35] found noise disturbance prevalence among seafarers of 15.6%, which is similar to the number of respondents in this study who considered leaving a ship due to noise (13.3%). The results show that noise pollution on board is not only temporary but permanent. Moreover, the research conducted by Szczepański and Otto [36] long ago found that noise levels during travel over and over exceed accepted norms, and reversible hearing impairment has been recorded after just one trip already.

Noise perception is an uncertain category. The estimated number of people exposed to noise is always lower than realistic. The number of exposed people who have disturbances due to noise exposure is uncertain as it is often a subjective assessment of an individual. Noise propagation from a single source is variable, while the spread from multiple sources is fraught with uncertainty. Noise protection measures can be primary, reducing noise at source and secondary such as noise propagation prevention, noise protection at home and workplace, economic measures, and regulations. They are individually very costly, and their effectiveness is generally low or uncertain [8]. However, Bowes et al. [37] showed that the costs of treatment and other compensation for hearing loss on navy ships are 15 times higher than investing in prevention programs, which offers, among other benefits, the possibility of significant savings.

5. Conclusions

Although increasingly supported by scientific evidence, the impact of noise on health has not yet been accepted as a real danger remaining underestimated without reaching full social awareness. Methods for external noise costs calculation remain subjective. The uncertainty of the noise nature and the limited motivation of the research community are reasons that little have been done to reduce noise in line with sustainable transport development. It is necessary to raise awareness of the damage caused by transport and its possible influence on the decision-making process in selecting the most appropriate transport mode. Education is crucial in raising awareness of noise detriment. The recent findings on the noise impact reveal greater exposure and more comprehensive health disorders than previously thought. This study contributes to raising awareness and the overall perception of noise pollution in maritime affairs, but with a small sample of seafarers, which cannot be considered representative, limits the results values. Within a surveyed period, seafarers underwent additional training, and their knowledge might be better than in the general population of seafarers. Furthermore, unlike the general population, this group is occupationally exposed to noise, and thus attitudes towards noise pollution are likely to be partly personally motivated. Limited perception and attitudes toward noise on board would probably be even more prominent by removing weaknesses from the research. Further research should include noise measurements inside the ship, which will provide correct noise exposure data to the workers and compare them with the noise perceived. It is also necessary to investigate the proportion of noise pollution topics in maritime education programs, aiming to increase the practical knowledge level and awareness of the noise impacts on health and society.

Author Contributions: Conceptualization, L.V.; methodology, L.V.; validation, V.M. and V.P.; formal analysis, V.M.; resources, V.P. and L.V.; data curation, L.F.; writing—original draft preparation, L.F.; writing—review and editing, L.V.; visualization, V.P.; supervision, V.M. and L.F. All authors have read and agreed to the published version of the manuscript.

Funding: This research received no external funding.

Institutional Review Board Statement: The study was conducted in accordance with the Declaration of Helsinki, and the protocol was approved by the Ethics Committee of the Faculty of Maritime Studies, University of Split (2181-197-01-06-0003).

Informed Consent Statement: Informed consent was obtained from all subjects involved in the study.

Data Availability Statement: The data presented in this study are available on request from the corresponding author.

Acknowledgments: The authors gratefully acknowledge the interviewed seafarers for participating in the research.

Conflicts of Interest: The authors declare no conflict of interest.

Abbreviations

α	critical *p* value
€	euro
CORREL	correlation test
dB	decibel, sound pressure unit
dB(A)	filter A—to measure on the hearing scale of a human ear
EC	European Community
EU	European Union
HP	hedonic price
IMO	International Maritime Organization
L_p	level of sound pressure
LR	linear regression test
m	meter
n	number
MI	monthly income
p-value	level of statistical significance
r	correlation coefficient in the correlation test
R^2	determination coefficient in the linear regression test
SOLAS	Convention for the Safety of Life at Sea
SPL	sound pressure level
WHO	World Health Organization
WTA	willingness to accept
WTP	willingness to pay
y	years

Appendix A

Questionnaire

Instructions—For multiple-choice questions, select only one and mark it with bold letters, color, or some other mark.

Noise is one of the biggest public health problems today. More than 20% of the population of the European Union is exposed to noise. Health problems due to noise pollution vary from annoyance and anxiety, concentration disturbances, and sleep disorders to the damage of the auditory organs damage, high blood pressure, and heart attack. Noise exposure causes anxiety in at least 13% of people. Traffic is the principal cause of environmental noise. Noise above 50 dB (intensity of a normal conversation in your home) is harmful to health, and above 65 dB (louder conversation in a cafe/restaurant) should not be tolerated. Noise intensity of 65 dB is 15 times higher than noise intensity of 50 dB. Individual procedures that subsequently install noise reduction elements are very costly, reducing volume by a maximum of 10 dB, and most often 2–3 dB.

Use the attached table to make it easier to estimate the intensity of the noise you are exposed to Appendix A (Table A1).

Table A1. Display of decibel level comparison [38].

Examples	Sound Pressure Level L_p dB SPL
Jet plane, 50 m distance	140
Pain threshold	130
Discomfort threshold	120
Chainsaw, 1 m distance	110
Disco club, 1 m distance from the speakers	100
Truck, 10 m distance	90
Rush hour road, 5 m distance	80
Vacuum cleaner, 1 m distance	70
Normal conversation, 1 m	60
Average house noise	50
Silent library	40
Bedroom at night	30
TV studio noise	20
Falling leaf	10
Hearing threshold	0

Table A2. General Data.

Year of Birth	
Marital status	Married Unmarried
Number of children	
Place of residence	City Village
Education level	Primary Secondary Bachelor Master
Profession	
Work experience	years
Type of work	
Monthly income	<2000 € 2000–3000 € 3000–4000 € 4000–5000 € >5000 €

Table A3. General Noise Perception.

Do you think that noise pollution is a significant environmental problem?	Yes No
Have you been aware of the harmful effects of noise on health so far?	Yes No Partly
To protect your health, how much money a year would you be willing to spend to reduce noise?	0 1–50 € 50–100 € 100–150 € 150–300 € >300 €
If you do not want to spend anything to reduce noise, explain why you would decide to do so	

Table A4. Noise on Board.

How much are you exposed to excessive noise on board?	Very much Much Little Very little Not at all
What exactly is the source of the noise that is disturbing you in your workplace?	
How much does the noise disturb you while you are resting or sleeping on ship?	Very much Much Little Very little Not at all
What exactly is the source of the noise that is disturbing you while you are resting?	
Are you exposed to vibration due to noise?	Yes No I do not know
Based on the attached decibel level comparison table, estimate how much noise intensity (in dB) you are exposed to on board: - in working hours - during rest	dB dB
How much does noise interfere with your work?	Very much Much Little Very little Not at all
How much does the noise distract you during your rest hours?	Very much Much Little Very little Not at all
Do you use noise protection equipment?	Always Sometimes Never
Have you ever considered leaving the ship due to noise?	Yes No I do not know

Table A4. *Cont.*

Do you think you should have a salary supplement due to noise?	Yes No I do not know
If the answer to the previous question is YES, what salary supplement (in percentage) do you think you should receive?	1% 5% 10% 15%
Explain why you chose that answer to the previous question?	
If you were thinking about getting off the ship what would be the reasons?	Just due to noise Due to noise and other reasons I would not go though I do not know
What do you think is the bigger environmental problem in maritime transport?	Air pollution Noise

Table A5. Noise in the Place of Residence.

Are you disturbed by outside noise in your apartment?	Very much Much Little Very little Not at all
When does it disturb you the most?	During the day At night During day and night Does not disturb at all I do not know
Based on the attached decibel level comparison table, estimate how much noise intensity you are exposed to in your apartment (in dB)?	During the day dB At night dB
Have you thought about moving because of the noise?	Yes No I do not know
How much do you think (in percentage) noise should affect the value of your apartment?	1% 5% 10% 15% >15%

References

1. Maibach, M.; Schreyer, C.; Sutter, D.; Van Essen, H.P.; Boon, B.H.; Smokers, R.; Schroten, A.; Doll, C.; Pawlowska, B.; Bak, M. *Handbook on Estimation of External Costs in the Transport Sector–IMPACT D1 Version 1.1*; INFRAS, Report Delft; CE Delft: Delft, The Netherlands, 2008.
2. Korzhenevych, A.; Dehnen, N.; Bröcker, J.; Holtkamp, M.; Meier, H.; Gibson, G.; Varma, A.; Cox, V. *Update of the Handbook on External Costs of Transport*; RICARDO-AEA: Oxfordshire, UK, 2014.
3. Van Essen, H.; Van Wijngaarden, L.; Schroten, A.; De Bruyn, S.; Sutter, D.; Bieler, C.; Maffii, S.; Brambilla, M.; Fiorello, D.; Fermi, F.; et al. *Handbook on the External Costs of Transport*; CE Delft: Delft, The Netherlands, 2019.
4. World Health Organization (WHO). *Burden of Disease from Environmental Noise-Quantification of Healthy Life Years Lost in Europe*; WHO Regional Office for Europe: Copenhagen, Denmark, 2011.
5. Andersson, H.; Jonsson, L.; Ögren, M. Benefit measures for noise abatement: Calculations for road and rail traffic noise. *Eur. Transp. Res. Rev.* **2013**, *5*, 135–148. [CrossRef]
6. European Commission. *Noise Impacts on Health, Science for Environment Policy*; European Commission Publications: Brussels, Belgium, 2015.

7. Peeters, B.; Van Blokland, G. *Decision and Cost/Benefit Methods for Noise Abatement Measures in Europe*; European Network of the Heads of Environment Protection Agencies (EPA Network): Vught, The Netherlands, 2018.
8. Starčević, S.M.; Bojović, N.J. Noise as an external effect of traffic and transportation. *Vojnoteh. Glas. Mil. Tech. Cour.* **2016**, *64*, 866–891. [CrossRef]
9. Fredianelli, L.; Carpita, S.; Licitra, G. A procedure for deriving wind turbine noise limits by taking into account annoyance. *Sci. Total Environ.* **2019**, *648*, 728–736. [CrossRef] [PubMed]
10. Badino, A.; Borelli, D.; Gaggero, T.; Rizzuto, E.; Schenonea, C. Noise emitted from ships; impact inside and outside the vessels. *Procedia Soc. Behav. Sci.* **2012**, *48*, 848–879. [CrossRef]
11. Oldenburg, M.; Felten, C.; Hedtmann, J.; Jensen, H.J. Physical influences on seafarers are different during their voyage episodes of port stay, river passage and sea passage: A maritime field study. *PLoS ONE* **2020**, *15*, e0231309. [CrossRef] [PubMed]
12. Picu, L.; Rusu, E.; Picu, M. An analysis of the noise in the engine room-case study a merchant ship navigating on Danube. In Proceedings of the 19th International Multidisciplinary Scientific GeoConference SGEM 2019, Albena, Bulgaria, 28 June–7 July 2019. [CrossRef]
13. Turan, O.; Helvacioglu, I.H.; Insel, M.; Khalid, H.; Kurt, R.E. Crew noise exposure on board ships and comparative study of applicable standards. *Ships Offshore Struct.* **2011**, *6*, 323–338. [CrossRef]
14. Mansi, F.; Cannone, E.S.S.; Caputi, A.; De Maria, L.; Lella, L.; Cavone, D.; Vimercati, L. Occupational Exposure on Board Fishing Vessels: Risk Assessments of Biomechanical Overload, Noise and Vibrations among Worker on Fishing Vessels in Southern Italy. *Environments* **2019**, *6*, 127. [CrossRef]
15. Sunde, E.; Bratveit, M.; Pallesen, S.; Moen, B.E. Noise and sleep on board vessels in the Royal Norwegian Navy. *Noise Health* **2016**, *18*, 85–92. [CrossRef] [PubMed]
16. Chia, R.; Tam, I.; Dev, A.K. Maritime Sustainability and Maritime Labour Convention-Reducing Vibration and Noise Levels on Board Ships for Health and Safety of Seafarers. In Proceedings of the MARTECH 2017 SINGAPORE Conference, Towards 2030: Maritime Sustainability through People and Technology, Singapore, 20–21 September 2017. Ultra Supplies.
17. Jégaden, D.; Lucas, D. About the relationship between ship noise and the occurrence of arterial hypertension in seafarers. *Int. Marit. Health* **2020**, *71*, 301. [CrossRef] [PubMed]
18. Kaerlev, L.; Jensen, A.; Nielsen, P.S.; Olsen, J.; Hannerz, H.; Tuchsen, F. Hospital contacts for noise-related hearing loss among Danish seafarers and fishermen: A population-based cohort study. *Noise Health* **2008**, *10*, 41–45. [CrossRef] [PubMed]
19. Norwegian Centre for Maritime Medicine. *Textbook of Maritime Medicine*, 2nd ed. 2013. Available online: http://textbook. maritimemedicine.com/ (accessed on 5 January 2021).
20. Nastasi, M.; Fredianelli, L.; Bernardini, M.; Teti, L.; Fidecaro, F.; Licitra, G. Parameters Affecting Noise Emitted by Ships Moving in Port Areas. *Sustainability* **2020**, *12*, 8742. [CrossRef]
21. Bolognese, M.; Fidecaro, F.; Palazzuoli, D.; Licitra, G. Port Noise and Complaints in the North Tyrrhenian Sea and Framework for Remediation. *Environments* **2020**, *7*, 17. [CrossRef]
22. Fredianelli, L.; Nastasi, M.; Bernardini, M.; Fidecaro, F.; Licitra, G. Pass-by Characterization of Noise Emitted by Different Categories of Seagoing Ships in Ports. *Sustainability* **2020**, *12*, 1740. [CrossRef]
23. Witte, R. *Regulation of Noise from Moored Ships, Green Port*; Mercator Media Ltd.: Fareham, Hampshire, UK, 2015; Available online: https://www.greenport.com/news101/Regulation-and-Policy/regulation-of-noise-from-moored-ships (accessed on 15 January 2021).
24. International Maritime Organization (IMO). Resolution MSC 338(91), Adoption of Amendments to the International Convention for the Safety of Life at Sea, 1974, as Amended, ANNEX 1 Resolution MSC 337(91) Adoption of the Code on Noise Levels on Board Ships (adopted on 30 November 2012). Available online: https://wwwcdn.imo.org/localresources/en/KnowledgeCentre/ IndexofIMOResolutions/Documents/MSC%20-%20Maritime%20Safety/337(91).pdf (accessed on 2 February 2021).
25. Bernardini, M.; Fredianelli, L.; Fidecaro, F.; Gagliardi, P.; Nastasi, M.; Licitra, G. Noise Assessment of Small Vessels for Action Planning in Canal Cities. *Environments* **2019**, *6*, 31. [CrossRef]
26. Alnuman, N.; Ghnimat, T. Awareness of Noise-Induced Hearing Loss and Use of Hearing Protection among Young Adults in Jordan. *Int. J. Environ. Res. Public Health* **2019**, *16*, 2961. [CrossRef] [PubMed]
27. European Commission. Minimum Health and Safety Requirements Regarding the Exposure of Workers to the Risks Arising from Physical Agents (Noise), Directive 2003/10/EC. Available online: https://eur-lex.europa.eu/legal-content/EN/TXT/PDF/?uri= CELEX:32003L0010&from=EN (accessed on 15 February 2021).
28. Park, J.; Chung, S.; Lee, J.; Sung, J.H.; Cho, S.W.; Sim, C.S. Noise sensitivity, rather than noise level, predicts the non-auditory effects of noise in community samples: A population-based survey. *BMC Public Health* **2017**, *17*, 315. [CrossRef] [PubMed]
29. Istamto, T.; Houthuijs, D.; Lebret, E. Willingness to pay to avoid health risks from road-traffic-related air pollution and noise across five countries. *Sci. Total Environ.* **2014**, *497–498*, 420–429. [CrossRef] [PubMed]
30. European Environmental Agency. *Environmental Noise in Europe–2020*; EEA Report No 22/2019; Publications Office of the European Union: Luxembourg, 2020. [CrossRef]
31. Health Canada. Guidance for Evaluating Human Health Impacts in Environmental Assessment: Noise, Health Canada, Ottawa. 2017. Available online: http://publications.gc.ca/site/eng/9.832514/publication.html (accessed on 3 May 2021).
32. Civil Aviation Authority. Aircraft Noise and Annoyance: Recent findings, No CAP 1588, Environmental Research and Consultancy Department. 2018. Available online: https://publicapps.caa.co.uk/modalapplication.aspx?appid=11&mode=detail&id=8246 (accessed on 3 May 2021).

33. Picu, L.; Rusu, E.; Picu, M. Quantification of vibration and noise transmitted to navigation personnel on a cargo ship on the Danube in a meta-analysis. In Proceedings of the 20th International Multidisciplinary Scientific GeoConference SGEM 2020, Albena, Bulgaria, 26 June–5 July 2021. [CrossRef]
34. Licitra, G.; Bolognese, M.; Palazzuoli, D.; Fredianelli, L.; Fidecaro, F. Port noise impact and citizens' complaints evaluation in rumble and mon acumen Interreg projects. In Proceedings of the 26th International Congress on Sound and Vibration, ICSV 2019, Montreal, QC, Canada, 7–11 July 2019.
35. Bernotaitė, L.; Malinauskienė, V. Joint effect of noise annoyance and workplace bullying on psychological distress among seafarers. In Proceedings of the 27th Conference of the International Society for Environmental Epidemiology "Addressing Environmental Health Inequalities, São Paulo, Brasil, 30 August–3 September 2015.
36. Szczepański, C.; Otto, B. Evaluation of exposure to noise in seafarers on several types of vessels in Polish Merchant Navy. *Bull. Inst. Marit Trop Med. Gdynia* **1995**, *46*, 13–17. [PubMed]
37. Bowes, M.; Shaw, G.; Trost, R.; Ye, M. *Computing the Return on Noise Reduction Investments in Navy Ships: A Life Cycle Cost Approach*; Mark Center Visitor Control Center: Alexandria, VA, USA, 2006.
38. Bilan, O. *Room Acoustics, Loudspeakers, Power Amplifiers and Speaker Cables*; Bilan: Split, Croatia, 1998; p. 500. Available online: http://www.audiologs.com/ozrenbilan/zvutlakrazint.htm (accessed on 21 January 2021).

MDPI

St. Alban-Anlage 66

4052 Basel

Switzerland

Tel. +41 61 683 77 34

Fax +41 61 302 89 18

www.mdpi.com

International Journal of Environmental Research and Public Health Editorial Office

E-mail: ijerph@mdpi.com

www.mdpi.com/journal/ijerph